BIOAVAILABILITY, BIOACCESSIBILITY AND MOBILITY OF ENVIRONMENTAL CONTAMINANTS

Analytical Techniques in the Sciences (AnTS)

Series Editor: David J. Ando, Consultant, Dartford, Kent, UK

A series of open learning/distance learning books which covers all of the major analytical techniques and their application in the most important areas of physical, life and materials sciences.

Titles Available in the Series

Analytical Instrumentation: Performance Characteristics and Quality
Graham Currell, University of the West of England, Bristol, UK

Fundamentals of Electroanalytical Chemistry
Paul M.S. Monk, Manchester Metropolitan University, Manchester, UK

Introduction to Environmental Analysis
Roger N. Reeve, University of Sunderland, UK

Polymer Analysis
Barbara H. Stuart, University of Technology, Sydney, Australia

Chemical Sensors and Biosensors
Brian R. Eggins, University of Ulster at Jordanstown, Northern Ireland, UK

Methods for Environmental Trace Analysis
John R. Dean, Northumbria University, Newcastle, UK

Liquid Chromatography–Mass Spectrometry: An Introduction
Robert E. Ardrey, University of Huddersfield, Huddersfield, UK

Analysis of Controlled Substances
Michael D. Cole, Anglia Polytechnic University, Cambridge, UK

Infrared Spectroscopy: Fundamentals and Applications
Barbara H. Stuart, University of Technology, Sydney, Australia

Practical Inductively Coupled Plasma Spectroscopy
John R. Dean, Northumbria University, Newcastle, UK

Bioavailability, Bioaccessibility and Mobility of Environmental Contaminants
John R. Dean, Northumbria University, Newcastle, UK

Forthcoming Titles

Quality Assurance in Analytical Chemistry
Elizabeth Prichard and Vicki Barwick, Laboratory of the Government Chemist, Teddington, UK

Techniques of Modern Organic Mass Spectrometry
Robert E. Ardrey, University of Huddersfield, Huddersfield, UK

The Analysis of DNA
Helen Hooper and Ruth Valentine, Northumbria University, Newcastle, UK

Analytical Profiling and Comparison of Drugs of Abuse
Michael D. Cole, Anglia Polytechnic University, Cambridge, UK

To my family

To Sam, my son, for keeping himself in his bedroom playing computer games and not disturbing me

To my daughter, Naomi, for her social and sporting activities that keep her busy

To the dog, Emmi (a border terrier), for keeping her barking under control

Finally, to my wife, Lynne, for her patience and understanding in keeping all those above under control while I spent hours in the study

Contents

Series Preface

There has been a rapid expansion in the provision of further education in recent years, which has brought with it the need to provide more flexible methods of teaching in order to satisfy the requirements of an increasingly more diverse type of student. In this respect, the *open learning* approach has proved to be a valuable and effective teaching method, in particular for those students who for a variety of reasons cannot pursue full-time traditional courses. As a result, John Wiley & Sons, Ltd first published the Analytical Chemistry by Open Learning (ACOL) series of textbooks in the late 1980s. This series, which covers all of the major analytical techniques, rapidly established itself as a valuable teaching resource, providing a convenient and flexible means of studying for those people who, on account of their individual circumstances, were not able to take advantage of more conventional methods of education in this particular subject area.

Following upon the success of the ACOL series, which by its very name is predominately concerned with Analytical *Chemistry*, the *Analytical Techniques in the Sciences* (AnTS) series of open learning texts has been introduced with the aim of providing a broader coverage of the many areas of science in which analytical techniques and methods are increasingly applied. With this in mind, the AnTS series of texts seeks to provide a range of books which covers not only the actual techniques themselves, but *also* those scientific disciplines which have a necessary requirement for analytical characterization methods.

Analytical instrumentation continues to increase in sophistication, and as a consequence, the range of materials that can now be almost routinely analysed has increased accordingly. Books in this series which are concerned with the *techniques* themselves reflect such advances in analytical instrumentation, while at the same time providing full and detailed discussions of the fundamental concepts and theories of the particular analytical method being considered. Such books cover a variety of techniques, including general instrumental analysis, spectroscopy, chromatography, electrophoresis, tandem techniques, electroanalytical methods, X-ray analysis and other significant topics. In addition, books in this series also consider the *application* of analytical techniques in areas such as environmental science, the life sciences, clinical analysis, food science, forensic analysis, pharmaceutical science, conservation and archaeology, polymer science and general solid-state materials science.

Written by experts in their own particular fields, the books are presented in an easy-to-read, user-friendly style, with each chapter including both learning objectives and summaries of the subject matter being covered. The progress of the reader can be assessed by the use of frequent self-assessment questions (SAQs) and discussion questions (DQs), along with their corresponding reinforcing or remedial responses, which appear regularly throughout the texts. The books are thus eminently suitable both for self-study applications and for forming the basis of industrial company in-house training schemes. Each text also contains a large amount of supplementary material, including bibliographies, lists of acronyms and abbreviations, and tables of SI Units and important physical constants, plus where appropriate, glossaries and references to literature and other sources.

It is therefore hoped that this present series of textbooks will prove to be a useful and valuable source of teaching material, both for individual students and for teachers of science courses.

Dave Ando
Dartford, UK

Preface

This book builds upon previous texts by the same author, namely *Extraction Methods for Environmental Analysis* (Wiley), which was published in 1998, and *Methods for Environmental Trace Analysis* (Wiley), published in 2003. The latter was part of the same series as is this book, i.e. 'Analytical Techniques in the Sciences' (AnTS). This new book expands upon the techniques described in the previous two books and focuses on their application to assess the bioavailability and bioaccessibility of metals and persistent organic pollutants in environmental samples. The book is arranged in eight chapters as follows.

Chapter 1 considers the link between human health and contaminated land. A particular focus of this chapter concerns the regulations placed on local authorities in England as a result of the Environmental Protection Act 1990: Part IIA. Examples of Soil Guideline Values (SGVs) are presented, along with their use for defining limits associated with different land-uses, including residential, allotments and commercial/industrial sites. An approach to assess contaminated soils relative to SGVs is given based on the Mean-Value and Maximum-Value Tests. Examples are provided which allow the reader to calculate values for these two tests.

Chapter 2 considers the different sample preparation techniques applied to solid and liquid samples for elemental analysis. In particular, the need to 'destroy' the matrix by heat and/or acid(s) for solid samples, whereas in the case of liquids the emphasis is on pre-concentration and/or separation of metals from solution. In addition, the different analytical techniques capable of performing quantitative analysis of metals/'metalloids' in environmental samples are reviewed. A unique aspect of this chapter is the undertaking of SWOT analyses for elemental analysis techniques to highlight the parameters that should be considered when choosing between the different techniques.

Chapter 3 evaluates the different sample preparation techniques applied to solid and liquid samples for persistent organic pollutant analysis. In particular, the need to liberate these pollutants from the matrix by heat and/or organic solvents for solid samples, whereas in the case of liquids the emphasis is on pre-concentration and/or separation of persistent organic pollutants from solution. In addition, the different analytical techniques capable of performing quantitative analysis of persistent organic pollutants in environmental samples are reviewed. As for Chapter 2, a further unique aspect of this chapter is the undertaking of SWOT analyses, in this case for persistent organic pollutant analysis techniques to highlight the parameters that should be considered when choosing between the different approaches.

Chapter 4 considers the different sample preparation techniques applied for single and sequential extraction of metals from soils and sediments. In particular, the use of the single extraction methods (based on acetic acid, ammonium nitrate, calcium chloride, ethylenediamine tetraacetic acid (EDTA), diethylenetriamine pentaacetic acid (DTPA) and sodium nitrate) for metal bioavailability studies. Also considered is the use of a procedure based on the sequential extraction of metals. In addition, the use of bioassays, in the form of earthworms and plants, for assessing metal bioavailability is reviewed. Finally, the role of Certified Reference Materials in single and sequential extractions of metals from soils and sediments is considered.

Chapter 5 evaluates the different sample preparation techniques for non-exhaustive extraction (cyclodextrin, supercritical fluid extraction, sub-critical water extraction, solid-phase microextraction and membrane separations) of persistent organic pollutants from soils and sediments. In addition, a mathematical approach to predict weak and strong solvents, based on the Hildebrand solubility parameter, is proposed. Finally, the use of bioassays, in the form of earthworms and plants, for assessing persistent organic pollutant bioavailability is reviewed.

Chapter 6 describes the different approaches for assessing the oral bioac-cessibility of metals and persistent organic pollutants from solid environmental samples. After a brief introduction to human physiology, the key components in the design of an *in vitro* gastrointestinal approach are discussed. Finally, examples of the use of *in vitro* gastrointestinal extraction for metals and persistent organic pollutants are presented.

Chapter 7 contains four case studies from the author's own laboratory. Case study one considers the uptake of metals (Cd, Cu, Mn, Pb and Zn) by plants (lettuce, spinach, radish and carrot) grown in compost under greenhouse con-ditions. Case study two considers the oral bioaccessibility of metals (Cd, Cu, Mn, Pb and Zn) from plants (lettuce, spinach, radish and carrot) grown on con-taminated compost using an *in vitro* gastrointestinal extraction approach. Case study three considers the uptake of persistent organic pollutants (α-endosulfan, β-endosulfan and endosulfan sulfate) by lettuce plants grown in compost. Finally, case study four considers the oral bioaccessibility of persistent organic pollutants (α-endosulfan, β-endosulfan and endosulfan sulfate) by lettuce plants grown in compost using an *in vitro* gastrointestinal extraction approach.

The final chapter (Chapter 8) provides examples of forms that could be used to record laboratory information at the time of carrying out experiments. Guidelines are first given for the recording of information associated with sample pre-treatment. Then, specific forms are provided for the recording of sample preparation details associated with environmental samples containing metals or persistent organic pollutants. Finally, guidelines are given for the recording of information associated with the analysis of metals and persistent organic pollutants. This chapter concludes with a resource section detailing lists of journals, books (general and specific) and Web addresses that will act to supplement this text.

John R. Dean
Northumbria University, Newcastle, UK

Acknowledgements

This present text includes material which has previously appeared in three of the author's earlier books, i.e. *Atomic Absorption and Plasma Spectroscopy* (ACOL Series, 1997), *Extraction Methods for Environmental Analysis* (1998) and *Methods for Environmental Trace Analysis* (AnTS Series, 2003), all published by John Wiley & Sons, Ltd. The author is grateful to the copyright holders for granting permission to reproduce figures and tables from his three earlier publications.

Marisa Intawongse is acknowledged for the preparation of Tables 6.2 and 6.3, while Dr Renli Ma is acknowledged for the preparation of Table 6.4. All three tables have been modified by the author from the original source.

The front cover shows a collection of young seedlings being grown in pots containing contaminated soil, under artificial light. After reaching maturity, the plants (and soil) were collected and their metal contents determined. Bioavailability was assessed in terms of uptake of metals from soil to plants, while oral bioaccessibility was determined by subjecting plant leaves or roots to *in vitro* gastrointestinal extraction. Further details can be found in Chapter 7. (Picture provided by John R. Dean, Northumbria University, Newcastle, UK.)

Acronyms, Abbreviations and Symbols

AAS	atomic absorption spectroscopy
ACS	American Chemical Society
AES	atomic emission spectroscopy
ANOVA	analysis of variance
ASE	accelerated solvent extraction
ASV	anodic stripping voltammetry
BIDS	Bath Information and Data Services
CI	chemical ionization
COSHH	Control of Substances Hazardous to Health
CVAAS	cold-vapour atomic absorption spectroscopy
CoV	coefficient of variation
CRM	Certified Reference Material
Da	dalton (atomic mass unit)
DAD	diode-array detection
DCM	dichloromethane
ECD	electron-capture detector
EDTA	ethylenediamine tetraacetic acid
EI	electron impact
emf	electromotive force
ETAAS	electrothermal atomic absorption spectroscopy
eV	electronvolt
FID	flame-ionization detector
FAAS	flame atomic absorption spectroscopy
FPD	flame photometric detector
GC	gas chromatography
GFAAS	graphite-furnace atomic absorption spectroscopy

HCL	hollow-cathode lamp
HPLC	high performance liquid chromatography
HyAAS	hydride-generation atomic absorption spectroscopy
IC	ion chromatography
ICP	inductively coupled plasma
ICP–AES	inductively coupled plasma–atomic emission spectroscopy
ICP–MS	inductively coupled plasma–mass spectrometry
id	internal diameter
IR	infrared
LC	liquid chromatography
LOD	limit of detection
LOQ	limit of quantitation
M	relative molecular mass
MAE	microwave-accelerated extraction
MDL	minimum detectable level
MS	mass spectrometry
MSD	mass-selective detector
PFE	pressurized fluid extraction
PMT	photomultiplier tube
ppb	parts per billion (10^9)
ppm	parts per million (10^6)
ppt	parts per thousand (10^3)
RF	radiofrequency
RSC	The Royal Society of Chemistry
RSD	relative standard deviation
SAX	strong anion exchange
SCX	strong cation exchange
SD	standard deviation
SFE	supercritical fluid extraction
SGV	Soil Guideline Value
SI (units)	Système International (d'Unitès) (International System of Units)
SRM	Standard Reference Material
URL	uniform resource locator
USEPA	United States Environmental Protection Agency
UV	ultraviolet
V	volt
W	watt
WWW	World Wide Web

c	speed of light; concentration
E	energy; electric field strength
f	(linear) frequency
I	electric current

m	mass
p	pressure
R	molar gas constant
s	sample standard deviation (unbiased)
t	time; Student factor
V	electric potential
z	ionic charge
λ	wavelength
ν	frequency (of radiation)
σ	measure of standard deviation
σ^2	variance

About the Author

John R. Dean, B.Sc., M.Sc., Ph.D., D.I.C., D.Sc., FRSC, C.Chem., C.Sci., Cert. Ed., Registered Analytical Chemist

John R. Dean took his first degree in Chemistry at the University of Manchester Institute of Science and Technology (UMIST), followed by an M.Sc. in Analytical Chemistry and Instrumentation at Loughborough University of Technology, and finally a Ph.D. and D.I.C. in Physical Chemistry at the Imperial College of Science and Technology (University of London). He then spent two years as a postdoctoral research fellow at the Food Science Laboratory of the Ministry of Agriculture, Fisheries and Food in Norwich, in conjunction with the Polytechnic of the South West in Plymouth. His work there was focused on the development of directly coupled high performance liquid chromatography and inductively coupled plasma–mass spectrometry methods for trace element speciation in foodstuffs. This was followed by a temporary lectureship in Inorganic Chemistry at Huddersfield Polytechnic. In 1988, he was appointed to a lectureship in Inorganic/Analytical Chemistry at Newcastle Polytechnic (now Northumbria University). This was followed by promotion to Senior Lecturer (1990), Reader (1994), Principal Lecturer (1998) and Associate Dean (Research) (2004). He was also awarded a personal chair in 2004. In 1998, he was awarded a D.Sc. (University of London) in Analytical and Environmental Science and was the recipient of the 23rd Society for Analytical Chemistry (SAC) Silver Medal in 1995. He has published extensively in analytical and environmental science. He is an active member of The Royal Society of Chemistry (RSC) Analytical Division, having served as a member of the atomic spectroscopy group for 15 years (10 as Honorary Secretary), as well as a past Chairman (1997–1999). He has served on the Analytical Division Council for three terms and is a former Vice-President (2002–2004), as well as a past-Chairman of the North-East Region of the RSC (2001–2003).

Chapter 1

Contaminated Land and the Link to Human Health

Learning Objectives

- To appreciate the environmental impact that people have had on the land on which we live.
- To be aware of the statutory responsibilities that local authorities in England have as a result of the Environmental Protection Act 1990: Part IIA.
- To appreciate the link between contaminated land and human health.
- To be aware of the UK Soil Guidelines Values for metals and persistent organic pollutants.
- To define the different land uses that are linked to the UK Soil Guidelines Values.
- To be able to define the following terms: exposure, intake dose, uptake dose and bioavailability.
- To be able to calculate the mean-test value for an environmental pollutant.
- To be able to calculate the maximum-test value for an environmental pollutant.

1.1 Introduction

The land on which we live is a precious resource that needs to be protected and conserved. However, the utilization of the land by humans has resulted in its contamination. When you travel around the country, you become aware of

---

Bioavailability, Bioaccessibility and Mobility of Environmental Contaminants   John R. Dean
© 2007 John Wiley & Sons, Ltd

the issues associated with past industrial activity, for example, the remnants of former tin mines in Cornwall, derelict mine workings in Yorkshire and former mills in Lancashire, as well as the clearing of former industrial sites to build new houses and offices. All households and commercial activities generate waste that is removed by waste-disposal engineers from our bins.

SAQ 1.1

Where does the waste from your household end up?

In England, the Contaminated Land Regulations (Environmental Protection Act 1990: Part IIA) came into force in April 2000. These regulations require local authorities in England to identify sites which fall under the definition of 'contaminated land' and provide suitable remediation options that agree with the 'suitable-for-use' approach. The statutory responsibilities on local authorities include the following:

• To inspect their areas to identify contaminated land.

• To prepare and serve notifications of contaminated land.

• To establish whether sites should be designated as 'special sites' and thus become the responsibility of the Environment Agency.

• To serve remediation notices where necessary.

• To undertake assessment of the best practicable remediation option and test for 'reasonableness'.

• To consult other parties, including the Environment Agency.

• To compile and maintain registers of contaminated land.

The inspection of areas of potential contaminated land by local authorities is an important first step in this process. As well as inspecting the land, it is necessary to identify whether it has a significant pollutant linkage between a source (i.e. the contaminant), a pathway (e.g. ingestion) and a receptor (e.g. a human being). This present book is concerned with these linkages and the approaches available to establish whether pollutants are bioavailable.

1.2 Soil Guideline Values

In the UK, Soil Guideline Values (SGVs) have been and are being developed for a range of metals/metalloids and organic contaminants in order to assess the risk to humans. In selecting the potential contaminants to be explored, two criteria have been used [1]:

(1) Contaminants should be commonly found on many sites and at concentrations likely to cause harm.

(2) Contaminants that show a potential risk to humans and/or have the potential to cause issues associated with natural waters, ecosystems or the integrities of buildings.

The lists of potential inorganic and organic contaminants identified for the assessment of industrial land in the UK (and their receptors) are shown in Table 1.1 and 1.2, respectively. Soil Guideline Values represent 'intervention values', i.e. an indicator that a soil concentration above the stated level might provide an

Table 1.1 Inorganic contaminants and their receptors [1]

Contaminant	Receptors			
	Humans	Water	Vegetation and ecosystem	Construction materials
Metals				
Barium		✓		
Beryllium	✓	✓	✓	
Cadmium	✓	✓	✓	
Chromium	✓	✓		
Copper		✓	✓	
Lead	✓	✓	✓	
Mercury	✓	✓	✓	
Nickel	✓	✓	✓	
Vanadium	✓	✓		
Zinc			✓	
'Semi-metals' and non-metals				
Arsenic	✓	✓		
Boron		✓	✓	
Selenium	✓	✓	✓	
Sulfur	✓		✓	✓
Inorganic chemicals				
Cyanide (complex)	✓	✓	✓	✓
Cyanide (free)[a]	✓	✓	✓	
Nitrate		✓		
Sulfate		✓	✓	✓
Sulfide		✓	✓	✓
Other				
Asbestos	✓			
pH (acidity/alkalinity)	✓	✓	✓	✓

[a]Free cyanide is broadly equivalent to 'easily liberatable cyanide', which covers compounds that can release hydrogen cyanide at pH 4 and 100°C.

Table 1.2 Organic contaminants and their receptors [1]

Contaminant	Receptors			
	Humans	Water	Vegetation and ecosystem	Construction materials
Acetone	✓	✓		
Oil/fuel hydrocarbons	✓	✓	✓	✓
Aromatic hydrocarbons				
Benzene	✓	✓	✓	✓
Chlorophenols	✓	✓	✓	✓
Ethylbenzene	✓	✓	✓	✓
Phenol	✓	✓	✓	✓
Toluene	✓	✓	✓	✓
o-Xylene	✓	✓	✓	✓
m,p-Xylene	✓	✓	✓	✓
Polycyclic aromatic hydrocarbons	✓	✓		
Chlorinated aliphatic hydrocarbons				
Chloroform	✓	✓	✓	
Carbon tetrachloride	✓	✓	✓	✓
Vinyl chloride	✓	✓		
1,2-Dichloroethane	✓	✓	✓	✓
1,1,1-Trichloroethane	✓	✓	✓	✓
Trichloroethene	✓	✓	✓	✓
Tetrachloroethene	✓	✓	✓	✓
Hexachlorobuta-1,3-diene	✓	✓	✓	
Hexachlorocyclohexanes	✓	✓	✓	
Dieldrin	✓	✓	✓	
Chlorinated aromatic hydrocarbons				
Chlorobenzenes	✓	✓	✓	
Chlorotoluenes	✓	✓	✓	
Pentachlorophenol	✓	✓	✓	
Polychlorinated biphenyls	✓	✓	✓	
Dioxins and furans	✓	✓	✓	
Organometallics				
Organolead compounds	✓		✓	
Organotin compounds	✓	✓		

unacceptable risk to humans and that further investigation and/or remediation is required. As well as the numerical values associated with the concentration of the contaminant, limits are indicated for a range of land-uses. The standard land-uses defined are residential, allotments and commercial/industrial. These land-uses are defined as follows [1–11].

1.2.1 Residential

People live in a wide variety of dwellings including, for example, detached, semi-detached and terraced properties, up to two storeys high. This land-use takes into account several different house designs, including buildings based on suspended floors and ground-bearing slabs. It assumes that residents have private gardens and/or access to community open spaces close to the home. Exposure has been estimated with and without a contribution from eating home-grown vegetables, which represents the key difference in potential exposure to contamination between those living in a house with a garden and those living in a house where no private garden area is available.

1.2.2 Allotments

These represent the provision of open space, commonly made by the local authority, for local people to grow fruit and vegetables for their own consumption. Typically, each plot is about one-fortieth of a hectare, with several plots to a site. Although some allotment holders may choose to keep animals, including rabbits, hens and ducks, potential exposure to contaminated meat and eggs has not been considered.

1.2.3 Commercial/Industrial

There are many different kinds of workplace and work-related activities. This land-use assumes that work takes place in a permanent single-storey building, factory or warehouse where employees spend most time indoors involved in office-based or relatively light physical work. This land-use is not designed to consider those sites involving 100% 'hard cover' (such as car parks), where the risks to the site-user are from ingestion or skin contact, because of the implausibility of such exposures arising while the constructed surface remains intact.

Soil Guideline Values, according to the different land-uses, for selected metals are shown in Table 1.3 while those for organic compounds are shown in Table 1.4. It is observed that while the concentrations differ for each metal/organic compound, due to their different toxicities, some common aspects are observed with respect to land-use.

DQ 1.1

Which type of land-use has the highest concentration?

Answer
The highest concentrations are observed for land to be used for commercial/industrial activities.

Table 1.3 Soil Guideline Values according to land-use for selected metals

Standard land-use	Soil Guideline Values (mg kg^{-1} dry weight soil)						
	Arsenic[a]	Cadmium[b]	Chromium[c]	Lead[d]	Inorganic mercury[e]	Nickel[f]	Selenium[g]
Residential with plant uptake	20	1–8[h]	130	450	8	50	35
Residential without plant uptake	20	30	200	450	15	75	260
Allotments	20	1–8[h]	130	450	8	50	35
Commercial/ industrial	500	1400	5000	750	480	5000	5000

[a] Data taken from DEFRA/EA [2].
[b] Data taken from DEFRA/EA [3].
[c] Data taken from DEFRA/EA [4].
[d] Data taken from DEFRA/EA [5].
[e] Data taken from DEFRA/EA [6].
[f] Data taken from DEFRA/EA [7].
[g] Data taken from DEFRA/EA [8].
[h] Values alter with respect to pH: 1, 2 and 8 mg kg^{-1} dry weight soil for pH 6, 7 and 8, respectively.

Table 1.4 Soil Guideline Values according to land-use for selected organic compounds[a]

Standard land-use	Soil Guideline Values (mg kg^{-1} dry weight soil)								
	Ethylbenzene[b]			Phenol[c]			Toluene[d]		
	1% SOM	2.5% SOM	5% SOM	1% SOM	2.5% SOM	5% SOM	1% SOM	2.5% SOM	5% SOM
Residential with plant uptake	9	21	41	78	150	280	3	7	14
Residential without plant uptake	16	41	80	21900	34400	37300	3	8	15
Allotments	18	43	85	80	155	280	31	73	14
Commercial/ industrial	48 000	48 000	48 000	21 900	43 000	78 100	150	350	680

[a] SOM, soil organic matter.
[b] Data taken from DEFRA/EA [10].
[c] Data taken from DEFRA/EA [11].
[d] Data taken from DEFRA/EA [9].

As can be seen above from the definition of commercial/industrial land, the risk to humans is minimal. This is in contrast to other land-uses, i.e. allotments and residential, where the SGVs are significantly lower.

DQ 1.2

What is the consequence to humans of low SGVs?

Answer

In these situations, the risk to humans is greatest and therefore the soil requirement in terms of metal concentration is lower.

It is also noted that cadmium (Table 1.3) has SGVs for 'residential-with-plant-uptake' sites and allotments which are pH-dependent. In addition, all data for organic compounds (Table 1.4) are soil organic matter-dependent.

1.3 Risk to Humans

The risk to humans from exposure to environmental contaminants is a major concern of regulatory agencies in countries worldwide. In this context, environmental contaminants relates to organic compounds, metals, metalloids, inorganic chemicals and others (including asbestos). Exposure to these environmental contaminants can result from a variety of pathways.

SAQ 1.2

Which types of *direct* exposure pathways can you think of?

DQ 1.3

Which types of *indirect* exposure pathways can you think of?

Answer

Indirect exposure can occur via pathways through the food chain, including contaminant uptake by plants followed by consumption by animals or humans. In the case of the former, exposure can continue through the food chain for the majority of humans who pursue an omnivorous or carnivorous diet.

Exposure can be defined as 'the amount of chemical that is available for intake by a target population at a particular site. Exposure is quantified as the concentration of the chemical in the medium (for example, air, water or food) integrated over the duration of exposure. It is expressed in terms of mass of substance per kg of soil, unit volume of air or litre of water (for example, $mg\,kg^{-1}$, $mg\,m^{-3}$ or $mg\,l^{-1}$)' [12]. As chemical exposure to humans occurs externally, this is referred to as the intake dose. The **intake dose** is defined as follows: 'the amount of a chemical entering or contacting the human body at the point of entry (that is, mouth, nose or skin) by ingestion, inhalation or skin contact. Actual intake will be a function of the chemical characteristics and the nature of the target

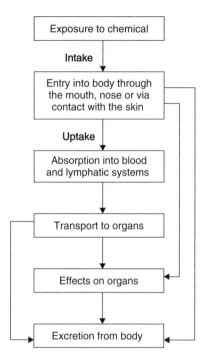

Figure 1.1 Human exposure to chemicals and the possible pathways (modified from DEFRA/EA [12]).

population and their behaviour patterns. Intake dose is expressed in terms of mass of substance per kg of body weight over a period of time (for example, $mg\,kg^{-1}$ $bw\,d^{-1}$)' [12]. However, not all substances that are 'intaken' by the body are absorbed and therefore it is possible to define the **uptake dose** as 'the amount of a contaminant that reaches the circulating blood having been absorbed by the body through the skin, the gastrointestinal system and the pulmonary system, expressed in terms of mass of substance per unit volume of blood (for example, $mg\,l^{-1}$)' [12]. In reality, the uptake dose is normally related to the intake (dose) by the **bioavailability** of the contaminant. Bioavailability is therefore defined as follows: 'the fraction of the chemical that can be absorbed by the body through the gastrointestinal system, the pulmonary system and the skin' [12]. The relationship between exposure, intake and uptake is shown in Figure 1.1.

1.4 An Approach to Assess Contaminated Soils Relative to Soil Guideline Values

The human health risk from a contaminated site can be determined after sampling of the designated area followed by the appropriate analytical technique to determine the level of contamination (see Chapters 2 and 3). Once the level of contamination, from the designated site, has been determined, the determined values can be compared to the appropriate SGVs. A statistical approach to compare values, has been determined [13] based on the **Mean-Value Test**. In order to compare values, the following information is required from the chosen, appropriately sampled, contaminated site:

- arithmetic sample mean
- standard deviation of the mean
- maximum concentration value obtained.

This approach is based on an identification of the 95% confidence limit of the measured (concentration) mean value and to compare the upper 95th percentile value with the SGV using the mean-value test described below.

1.4.1 Mean-Value Test [13]

The sample mean value (x), based on only a few samples, may be a poor estimate of the true (population) mean. Therefore, a 'no-remedial action' decision based on x less than the Soil Guideline Value, G, may not be adequately 'health-protective' when x is computed from only a small number of samples. It is therefore desirable to state with a given confidence (e.g. the 95th percentile) that the population mean is less then the value of G.

The steps involved in the calculation are as follows:

(1) Calculate the arithmetic sample mean, x.

(2) Calculate the (unbiased) sample standard deviation, s.

(3) Select an appropriate t-value from standard tables (Table 1.5 gives the t-values for a 95th percentile confidence limit).

Table 1.5 Single-tailed t-values (95th percentile confidence limits) and their relationships to sample sizes (n) [13]

n	t	n	t	n	t
—	—	11	1.812	21	1.725
2	6.314	12	1.796	22	1.721
3	2.920	13	1.782	23	1.717
4	2.353	14	1.771	24	1.714
5	2.132	15	1.761	25	1.711
6	2.015	16	1.753	26	1.708
7	1.943	17	1.746	27	1.706
8	1.895	18	1.740	28	1.703
9	1.860	19	1.734	29	1.701
10	1.833	20	1.729	30	1.699

(4) Calculate the upper 95th percentile bound of the sample as:

$$US_{95} = \bar{x} + \frac{ts}{\sqrt{n}}$$

(5) Compare the upper bound value (US_{95}) with the Soil Guideline Value, G.

If the upper bound value (US_{95}) is less than G, then the Mean-Value Test has been passed and the site may be considered not to present a significant possibility of significant harm to human health in the context of the regulations (Environmental Protection Act 1990: Part IIA). Conversely, if the test fails, then it is necessary to consider whether more samples should be taken (as the number of original samples for the test was low) or to make a determination as contaminated land (Environmental Protection Act 1990: Part IIA), taking into account the other requirements of the regime, such as presence of a significant pollutant linkage.

NOTE: Some analytical data sets will include samples determined as below the detection limit of the instrument used. In these cases, and for the purposes of performing the calculation, assign a value equal to the detection limit. For example, if the value reported for Pb is $< 0.5\,\mathrm{mg\,kg^{-1}}$, a value of $0.5\,\mathrm{mg\,kg^{-1}}$ should be used for calculation.

DQ 1.4

Carry out the Mean-Value Test [13] for the following ten samples taken from a residential site (without plant uptake) where the level of chromium contamination ($\mathrm{mg\,kg^{-1}}$ dry weight soil) was determined to be as follows:

$x_1 = 69$	$x_2 = 120$	$x_3 = 220$	$x_4 = 376$	$x_5 = 156$
$x_6 = 109$	$x_7 = 205$	$x_8 = 320$	$x_9 = 96$	$x_{10} = 196$

The Soil Guideline Value (G) for chromium is 200 (Table 1.3).

Answer
By using the above data, it is possible to:

(1) Calculate the arithmetic sample mean, $\bar{x}$, as 186.70.

(2) Calculate the (unbiased) sample standard deviation, s, as 99.24.

(3) Obtain the t-value from Table 1.5 for $n = 10$, as $t = 1.833$.

(4) Calculate the upper 95th percentile bound of the sample as:

$$US_{95} = \bar{x} + \frac{ts}{\sqrt{n}}$$

$$US_{95} = 186.70 + [(1.833)(99.24)]/\sqrt{10}$$

$$US_{95} = 244.22$$

(5) Compare the upper bound value (US_{95}) with the Soil Guideline Value, G.

The upper bound value is greater than the Soil Guideline Value of 200 and therefore the data do NOT pass the Mean-Value Test. In these circumstances, two options are possible: take further samples to gain a more representative sampling of the site or take remedial action, if, for example, further sampling is not practicable, or time-scales dictate rapid action.

SAQ 1.3

Carry out the Mean-Value Test for the following data and recommend what, if any, further action is required.
    The following twelve samples were taken from a commercial site and the level of arsenic contamination (mg kg^{-1} dry weight soil) was determined to be as follows:

$x_1 = 189$	$x_2 = 520$	$x_3 = 420$	$x_4 = 324$	$x_5 = 356$	$x_6 = 450$
$x_7 = 506$	$x_8 = 410$	$x_9 = 380$	$x_{10} = 389$	$x_{11} = 456$	$x_{12} = 502$

The Soil Guideline Value (G) for arsenic is 500.

1.4.2 Maximum-Value Test [13]

Individual contamination values that exceed the SGVs will always require further investigation, even when the Mean-Value Test has been passed. However, the issue of accepting or rejecting maximum values that exceed the SGVs is not straightforward. The primary goal of health protection needs to be balanced against the fact that sampling of contaminated sites, followed by analytical measurement, can lead to significant variability in the data obtained, particularly at low analyte concentrations. In deciding whether to take additional samples to analyse (as indicated in the example above) it is necessary to consider whether the maximum value in a given data set has come from the sample statistical population or indeed whether it is, in fact, a statistical outlier.

    One approach [13] assumes that all data are derived from a normal statistical population. It is argued [13] that it is reasonable to 'log' the measured values as this results in a more-or-less symmetric distribution which approximates to a normal distribution.

    In operation, the individual measurements ($x_1, x_2 \ldots x_n$) are 'log-transformed' prior to calculation of the sample mean ($\bar{y}$) and unbiased sample standard deviation (S_y). The Maximum-Value Test parameter (T) is therefore:

$$T = (y_{\max} - \bar{Y})/S_y$$

If the value of T is smaller than some critical value, the maximum value may be accepted (at the particular level of confidence) as a member of the underlying population from which the other measurements were taken. If T is greater than the critical value, the maximum value is treated as an *outlier*, which may indicate a localized area of contamination.

DQ 1.5

Carry out the Maximum-Value Test [13] for the following ten samples taken from a residential site (without plant uptake) where the level of chromium contamination (mg kg^{-1} dry weight soil) was determined to be as shown in Table DQ 1.5 (expressed in concentrations and log concentrations).

Table DQ 1.5 Sample data for the Maximum-Value Test

i	x_i	$y_i = \log x_i$
1	69	1.839
2	120	2.079
3	220	2.342
4	376	2.575
5	156	2.193
6	109	2.037
7	205	2.312
8	320	2.505
9	96	1.982
10	196	2.292

The objective is to decide whether the maximum value, $x_4 = 376$ ($y_{max} = 2.575$), should be treated as an outlier, or whether it is reasonable to be considered as coming from the same underlying population as the other samples.

Answer

By using the above data, it is possible to:

(1) Calculate the arithmetic sample mean of the y-values, as $\bar{y} = 2.216$.

(2) Calculate the (unbiased) standard deviation of the y values, as $s_y = 0.2337$.

Table 1.6 Critical values to test for the presence of outliers [13]

N	5%	10%
4	1.46	1.42
5	1.67	1.60
6	1.82	1.73
7	1.94	1.83
8	2.03	1.91
9	2.11	1.98
10	2.18	2.04
12	2.29	2.13
14	2.37	2.21
16	2.44	2.28
18	2.50	2.33
20	2.56	2.38

(3) Obtain the outlier test statistic as:

$$T = (y_{max} - \overline{Y})/s_y$$
$$T = (2.575 - 2.216)/0.2337 = 1.54$$

Then, compare this value ($T = 1.54$) with the critical value (T_{crit}) in Table 1.6.

In the context of critical values being used for contaminated soil analysis, the concern is with regard to accepting a value which should be treated as an outlier. As 10% critical values are more stringent, they should be used for health-protection purposes, rather than the 5% values.

Therefore, the maximum value statistic calculated above ($T = 1.54$) is less than the 10% critical value of 2.04. Thus, it is reasonable to conclude that the maximum value is indeed part of the same underlying distribution as the other values.

NOTE: It is worth remembering that passing an outlier test does not prove that a maximum value is not an outlier possibly representing an (undiscovered) area of contamination. It simply indicates that the maximum value is reasonably consistent with belonging to the same underlying distribution as other measurements.

SAQ 1.4

Carry out the Maximum-Value Test for the following data and determine whether the value is an outlier or not [13].

The following twelve samples were taken from a commercial site and the level of arsenic contamination (mg kg^{-1} dry weight soil) was determined to be as shown in Table SAQ 1.4 (expressed in concentrations and log concentrations).

Table SAQ 1.4 Sample data for the Maximum-Value Test

i	x_i	$y_i = \log x_i$
1	189	2.276
2	520	2.716
3	420	2.623
4	324	2.510
5	356	2.551
6	450	2.653
7	506	2.704
8	410	2.613
9	380	2.580
10	389	2.590
11	476	2.678
12	502	2.701

Summary

The link between human health and contaminated land is demonstrated in this chapter, in particular, the regulations placed on local authorities in the UK as a result of the Environmental Protection Act 1990: Part IIA. Examples of Soil Guideline Values (SGVs) are presented, along with their use for defining limits associated with different land-uses, including residential, allotments and commercial/industrial sites. An approach to assess contaminated soils relative to the SGVs is given based on the Mean-Value and Maximum-Value Tests.

References

1. DEFRA/EA (Department for Environment, Food and Rural Affairs and the Environment Agency), *Potential Contaminants for the Assessment of Land*, R&D Publication CLR 8, Environment Agency, UK, 2002 [http://www.environment-agency.gov.uk/].
2. DEFRA/EA (Department for Environment, Food and Rural Affairs and the Environment Agency), *Soil Guidelines Values for Arsenic Contamination*, R&D Publication SGV 1, Environment Agency, UK, 2002 [http://www.environment-agency.gov.uk/].
3. DEFRA/EA (Department for Environment, Food and Rural Affairs and the Environment Agency), *Soil Guidelines Values for Cadmium Contamination*, R&D Publication SGV 3, Environment Agency, UK, 2002 [http://www.environment-agency.gov.uk/].

4. DEFRA/EA (Department for Environment, Food and Rural Affairs and the Environment Agency), *Soil Guidelines Values for Chromium Contamination*, R&D Publication SGV 4, Environment Agency, UK, 2002 [http://www.environment-agency.gov.uk/].

5. DEFRA/EA (Department for Environment, Food and Rural Affairs and the Environment Agency), *Soil Guidelines Values for Lead Contamination*, R&D Publication SGV 10, Environment Agency, UK, 2002 [http://www.environment-agency.gov.uk/].

6. DEFRA/EA (Department for Environment, Food and Rural Affairs and the Environment Agency), *Soil Guidelines Values for Inorganic Mercury Contamination*, R&D Publication SGV 5, Environment Agency, UK, 2002 [http://www.environment-agency.gov.uk/].

7. DEFRA/EA (Department for Environment, Food and Rural Affairs and the Environment Agency), *Soil Guidelines Values for Nickel Contamination*, R&D Publication SGV 7, Environment Agency, UK, 2002 [http://www.environment-agency.gov.uk/].

8. DEFRA/EA (Department for Environment, Food and Rural Affairs and the Environment Agency), *Soil Guidelines Values for Selenium Contamination*, R&D Publication SGV 9, Environment Agency, UK, 2002 [http://www.environment-agency.gov.uk/].

9. DEFRA/EA (Department for Environment, Food and Rural Affairs and the Environment Agency), *Soil Guidelines Values for Toluene Contamination*, Science Report SGV 15, Environment Agency, UK, 2004 [http://www.environment-agency.gov.uk/].

10. DEFRA/EA (Department for Environment, Food and Rural Affairs and the Environment Agency), *Soil Guidelines Values for Ethylbenzene Contamination*, Science Report SGV 16, Environment Agency, UK, 2005 [http://www.environment-agency.gov.uk/].

11. DEFRA/EA (Department for Environment, Food and Rural Affairs and the Environment Agency), *Soil Guidelines Values for Phenol Contamination*, Science Report SGV 8, Environment Agency, UK, 2005 [http://www.environment-agency.gov.uk/].

12. DEFRA/EA (Department for Environment, Food and Rural Affairs and the Environment Agency), *Contaminants in Soil: Collation of Toxicological Data and Intake Values for Humans*, R&D Publication CLR 9, Environment Agency, UK, 2002 [http://www.environment-agency.gov.uk/].

13. DEFRA/EA (Department for Environment, Food and Rural Affairs and the Environment Agency), *Assessment of Risks to Human Health from Land Contamination: An Overview of the Development of Soil Guideline Values and Related Research*, R&D Publication CLR 7, Environment Agency, UK, 2002 [http://www.environment-agency.gov.uk/].

Chapter 2

Sample Preparation and Analytical Techniques for Elemental Analysis of Environmental Contaminants

Learning Objectives

- To appreciate the different approaches available for preparation of solid and liquid samples for elemental analysis.
- To be aware of the procedure for preparing a solid sample using acid digestion with a hot-plate and microwave oven.
- To be aware of other decomposition methods, i.e. fusion and dry ashing.
- To be aware of the procedure for preparing a liquid sample using liquid–liquid extraction.
- To be aware of other extraction methods, i.e. ion-exchange and co-precipitation.
- To understand and explain the principles of operation of atomic absorption spectroscopy (AAS).
- To describe the benefits and mode of operation of a flame, graphite furnace, hydride generation and cold-vapour generation as used in AAS.
- To understand and explain the principles of operation of an inductively coupled plasma.
- To understand the principles of operation of a range of nebulizers.

Bioavailability, Bioaccessibility and Mobility of Environmental Contaminants   John R. Dean
© 2007 John Wiley & Sons, Ltd

- To be able to describe the operation of an inductively coupled plasma–mass spectrometer.
- To appreciate the range of mass spectrometers used in inductively coupled plasma–mass spectrometry (ICP–MS).
- To be able to identify isobaric and molecular interferences in ICP–MS.
- To understand the role that a collision/reaction cell can play in the alleviation of interferences in ICP–MS.
- To understand and explain the principles of X-ray fluorescence spectroscopy.
- To understand the principles of anodic stripping voltammetry.
- To appreciate the concept of a hyphenated technique as associated with chromatography and atomic spectroscopy.

2.1 Introduction

The analysis of elements (metal, metalloids and some non-metals) in the Periodic Table can be carried out by using a variety of analytical techniques, including those based on atomic spectroscopy, X-ray fluorescence spectroscopy, mass spectrometry and electrochemistry.

SAQ 2.1

How do you think you would decide which analytical technique to use?

This section contains brief reviews on the principle techniques of atomic spectroscopy, including atomic absorption and atomic emission spectroscopy, energy-dispersive X-ray fluorescence spectrometry, inorganic mass spectrometry (principally inductively coupled plasma–mass spectrometry) and electrochemistry (anodic stripping voltammetry). However, first it is necessary to consider the different sample preparation approaches that may be required prior to analysis.

2.2 Sample Preparation for Elemental Analysis

2.2.1 Solid Samples

Solid samples for elemental analysis frequently require decomposition of the matrix to release the metals. An exemption to this is when analytical techniques are used to analyse solids directly (see, for example, Section 2.6 on X-ray fluorescence (XRF) spectroscopy). Decomposition involves the release of the metal from a matrix using a reagent (mineral/oxidizing acids or fusion flux) and/or heat.

2.2.1.1 Dry Ashing

This involves heating the sample in a silica or porcelain crucible in a muffle furnace in the presence of air at 400–800°C. After decomposition, the residue is dissolved in mineral acid and transferred to a volumetric flask prior to analysis.

DQ 2.1

Can you identify any problems that may arise when heating samples to high temperatures (400–800°C)?

Answer

Problems can arise due to loss of volatile elements, e.g. Hg and As. Therefore, while compounds can be added to retard the loss of volatiles its use is limited.

2.2.1.2 Acid Digestion (including Microwave)

Acid digestion involves the use of mineral or oxidizing acids and an external heat source to decompose the sample. A summary of the most common types of acids used and their applications is shown in Table 2.1. Acid digestion can be carried out in open glass vessels (beakers or boiling tubes) by using a hot-plate or multiple-sample digester (Figure 2.1).

DQ 2.2

What advantage does the use of a multiple-sample digester have over a hot-plate?

Answer

A multiple-sample digester allows a number of samples to be prepared simultaneously.

An example of a method for the acid digestion of soils is show in Figure 2.2.
An alternative approach to conventional heating involves the use of microwave heating.

SAQ 2.2

How does microwave heating work?

Commercial microwave systems allow multiple vessels to be heated simultaneously. The microwave energy output of a typical system may be 1500 W at a frequency of 2450 MHz at 100% power. Pressures (up to 800 psi) at temperatures

Table 2.1 Some examples of common acids used for wet decomposition. From Dean, J. R., *Atomic Absorption and Plasma Spectroscopy*, 2nd Edition, ACOL Series, John Wiley & Sons, Limited, Chichester, UK, 1997. © University of Greenwich and reproduced by permission of the University of Greenwich

Acid (s)	Boiling point (°C)	Comments
Hydrochloric (HCl)	110	Useful for salts of carbonates, phosphates, some oxides and some sulfides. A weak reducing agent; not generally used to dissolve organic matter
Sulfuric (H_2SO_4)	338	Useful for releasing a volatile product; good oxidizing properties for ores, metals, alloys, oxides and hydroxides; often used in combination with HNO_3. **Caution**: H_2SO_4 must never be used in PTFE vessels (PTFE has a melting point of 327°C and deforms at 260°C)
Nitric (HNO_3)	122	Oxidizing attack on many samples not dissolved by HCl; liberates trace elements as the soluble nitrate salt. Useful for dissolution of metals, alloys and biological samples
Perchloric ($HClO_4$)	203	At fuming temperatures, a strong oxidizing agent for organic matter. **Caution**: violent, explosive reactions may occur – care is needed. Samples are normally pre-treated with HNO_3 prior to addition of $HClO_4$
Hydrofluoric (HF)	112	For digestion of silica-based materials; forms SiF_6^{2-} in acid solution; **caution** is required in its use; glass containers should not be used, only plastic vessels. In case of spillages, calcium gluconate gel (for treatment of skin-contact sites) should be available prior to usage; evacuate to hospital immediately if skin is exposed to liquid HF
Aqua regia (nitric/hydrochloric)	—	A 1:3 vol/vol mixture of HNO_3:HCl is called aqua regia; forms a reactive intermediate, NOCl. Used for metals, alloys, sulfides and other ores – best known because of its ability to dissolve Au, Pd and Pt

[a] Protective clothing/eyewear is essential in the use of concentrated acids. All acids should be handled with care and in a fumecupboard.

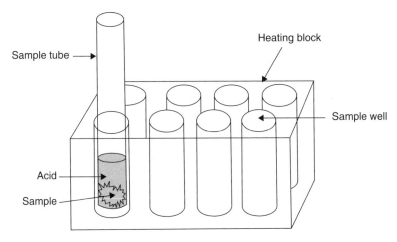

Figure 2.1 Schematic of a commercial acid digestion system. From Dean, J. R., *Methods for Environmental Trace Analysis*, AnTS Series. Copyright 2003. © John Wiley & Sons, Limited. Reproduced with permission.

(up to 300°C) are achievable. A typical operating procedure for the digestion of a sample of soil is shown in Figure 2.3.

2.2.1.3 Fusion

Occasionally, samples, e.g. silicates and oxides, are not destroyed by the action of acid. In this situation, fusion is used. In this process, excess reagent (e.g. sodium carbonate, lithium meta- or tetraborate and potassium pyrosulfate) and the sample are placed in a metal crucible, e.g. platinum, and heated in a muffle furnace (300–1000°C) for 1–2 h. After cooling, the melt can be dissolved in a mineral acid. The high salt content of the final solution may lead to problems in the subsequent analysis.

2.2.2 Liquid Samples

Normally, in the analysis of liquid samples for trace, minor and major metals, no sample pre-treatment apart from filtration (0.2 μm) is required. However, sample pre-treatment in the form of separation and/or pre-concentration may be required.

DQ 2.3

Why might pre-concentration of a sample be required?

Answer

This may be required if the chosen analytical technique is not sensitive enough to measure (ultra)-trace metal concentrations.

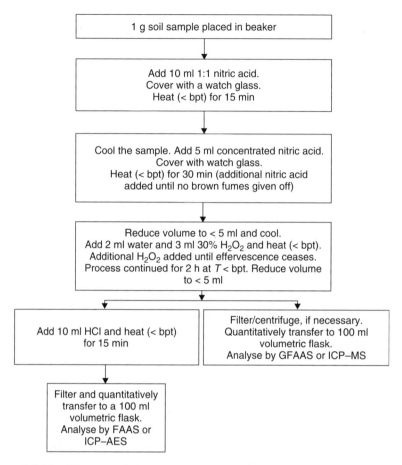

Figure 2.2 The EPA procedure for the acid digestion of sediments, sludges and soils using a hot-plate: GFAAS, graphite-furnace atomic absorption spectroscopy; FAAS, flame atomic absorption spectroscopy; ICP–MS, inductively coupled plasma–mass spectrometry; ICP–AES, inductively coupled plasma–atomic emission spectroscopy. From Dean, J. R., *Methods for Environmental Trace Analysis*, AnTS Series. Copyright 2003. © John Wiley & Sons, Limited. Reproduced with permission.

2.2.2.1 Liquid–Liquid Extraction

In this technique, a large volume of the aqueous sample, containing the metal, is mixed with a suitable chelating agent. After metal-complexation, the resultant species is collected in an immiscible organic solvent.

SAQ 2.3

What is a metal complex?

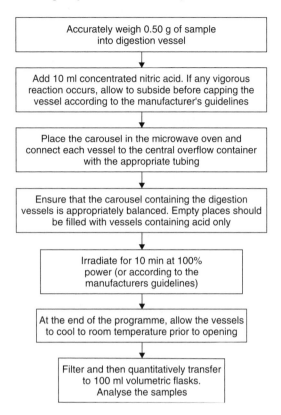

Figure 2.3 A typical procedure used for the microwave digestion of sediments, sludges and soils. From Dean, J. R., *Methods for Environmental Trace Analysis*, AnTS Series. Copyright 2003. © John Wiley & Sons, Limited. Reproduced with permission.

The basis of this technique is that the metal present in a large volume of sample water is effectively and quantitatively transferred into a small volume of organic solvent. The most commonly used chelate in atomic spectroscopy is ammonium pyrrolidine dithiocarbamate (APDC). This reagent, normally used as a 1–2% aqueous solution, is extracted with the organic solvent methylisobutyl ketone (MIBK). The procedure for the extraction of metals from solution using APDC and MIBK is shown in Figure 2.4.

2.2.2.2 Ion-Exchange

Cation-exchange chromatography is used to separate metal ions (positively charged species) from aqueous solution as follows:

(1) Metal ion (M^{n+}) pre-concentrated on cation-exchange resin

$$n\mathrm{RSO_3^-H^+} + \mathbf{M^{n+}} = \mathbf{(RSO_3^-)}n\mathbf{M^{n+}} + \mathrm{H^+} \qquad (2.1)$$

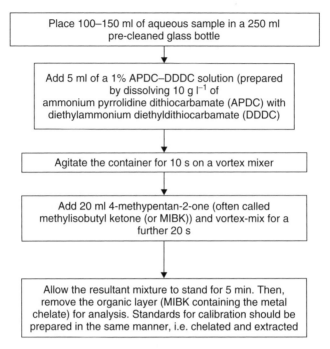

Place 100–150 ml of aqueous sample in a 250 ml
pre-cleaned glass bottle

Add 5 ml of a 1% APDC–DDDC solution (prepared
by dissolving 10 g l^{-1} of
ammonium pyrrolidine dithiocarbamate (APDC) with
diethylammonium diethyldithiocarbamate (DDDC)

Agitate the container for 10 s on a vortex mixer

Add 20 ml 4-methypentan-2-one (often called
methylisobutyl ketone (or MIBK)) and vortex-mix for a
further 20 s

Allow the resultant mixture to stand for 5 min. Then,
remove the organic layer (MIBK containing the metal
chelate) for analysis. Standards for calibration should be
prepared in the same manner, i.e. chelated and extracted

Figure 2.4 A typical procedure used for the liquid–liquid extraction of metals. From Dean, J. R., *Methods for Environmental Trace Analysis*, AnTS Series. Copyright 2003. © John Wiley & Sons, Limited. Reproduced with permission.

(2) Desorption of metal ion using acid

$$(\mathbf{RSO_3}^-)n\,\mathbf{M}^{n+} + \mathbf{H}^+ = n\mathrm{RSO_3}^-\mathrm{H}^+ + \mathbf{M}^{n+} \qquad (2.2)$$

An alternative approach, which allows the separation of an excess of alkali metal ions from other cations, uses a chelating ion-exchange resin (e.g. Chelex 100). This type of resin forms chelates with the metal ions. It has been found that Chelex-100, in acetate buffer at pH 5–6, can retain Al, Bi, Cd, Co, Cu, Fe, Ni, Pb, Mn, Mo, Sc, Sn, Th, U, V, W, Zn, Y and rare-earth metals, while at the same time not retaining alkali metals, alkali-earth metals and anions (F$^-$, Cl$^-$, Br$^-$ and I$^-$).

DQ 2.4

Identify the alkali metals.

Answer

Lithium (Li), sodium (Na), potassium (K), rubidium (Rb) and caesium (Cs).

DQ 2.5

Identify the alkali-earth metals.

Answer

Beryllium (Be), calcium (Ca), magnesium (Mg), strontium (Sr) and barium (Ba).

2.2.2.3 Co-Precipitation

This allows quantitative precipitation of the metal ion of interest by the addition of a co-precipitant, e.g. $Fe(OH)_3$, to the sample solution. Further sample preparation, e.g. dissolution, is required for the analyte prior to analysis.

2.3 Atomic Absorption Spectroscopy

Atomic absorption spectroscopy (AAS) is a commonly encountered technique in the laboratory due to its simplicity and low capital cost. Its limitations are based on its ability to be able to measure only one metal/metalloid at a time and its short linear dynamic range. The main components of an atomic absorption spectrometer are a radiation source (typically a hollow-cathode lamp), an atomization cell (typically a flame or graphite furnace) and methods of wavelength selection (monochromator) and detection (photomultiplier tube) (Figure 2.5). The hollow-cathode lamp (Figure 2.6) generates a characteristic narrow-line emission of a selected metal.

DQ 2.6

If you were going to analyse a sample for its iron content what hollow-cathode lamp would you need to have in the AAS instrument?

Answer

After considering the brief information provided so far, it should have become obvious that as the technique can only analyse one element at

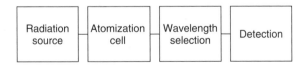

Figure 2.5 Block diagram of a typical atomic absorption spectrometer layout. From Dean, J. R., *Atomic Absorption and Plasma Spectroscopy*, 2nd Edition, ACOL Series, John Wiley & Sons, Limited, Chichester, UK, 1997. © University of Greenwich and reproduced by permission of the University of Greenwich.

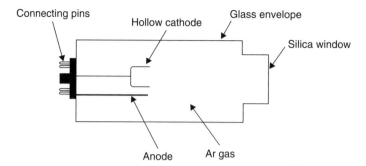

Figure 2.6 Schematic diagram of a hollow-cathode lamp used in atomic absorption spectroscopy. From Dean, J. R., *Atomic Absorption and Plasma Spectroscopy*, 2nd Edition, ACOL Series, John Wiley & Sons, Limited, Chichester, UK, 1997. © University of Greenwich and reproduced by permission of the University of Greenwich.

a time the choice of hollow-cathode lamp is important. Therefore, to analyse for iron requires an iron hollow-cathode lamp.

The atomization cell is the site were the sample (liquid or vapour) is introduced and dissociated, such that metal atoms are liberated from a hot environment. Typical atomization cells are a flame or graphite furnace. The hot environment of the atomization cell is sufficient to cause a broadening of the absorption line of the metal of interest. Utilizing the narrowness of the emission line from the hollow-cathode lamp (HCL), together with the broad absorption line created in the atomization cell, means that the monochromator only has to isolate the line of interest from other lines emitted by the HCL (Figure 2.7). This unique feature of

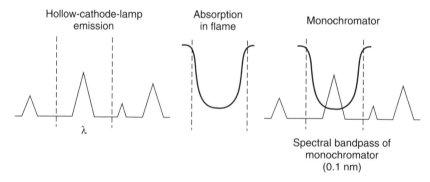

Figure 2.7 Basic principle of atomic absorption spectroscopy – the 'lock and key' effect. From Dean, J. R., *Atomic Absorption and Plasma Spectroscopy*, 2nd Edition, ACOL Series, John Wiley & Sons, Limited, Chichester, UK, 1997. © University of Greenwich and reproduced by permission of the University of Greenwich.

AAS gives it a high degree of selectivity, i.e. the ability to measure the element selected in preference to other elements present in the sample.

The most common atomization cell is the pre-mixed laminar flame in which the fuel and oxidant (air) gases are mixed, in an expansion chamber, prior to entering the slot-burner (the point of ignition of the flame). Two flames are used in AAS, i.e. the air–acetylene (temperature, 2500 K) or the nitrous oxide–acetylene (temperature, 3150 K) flame.

SAQ 2.4

What other flames could be used?

Both are located in a slot-burner which is positioned in the light path of the HCL (Figure 2.8). The air–acetylene flame (slot length, 100 mm) is the most commonly used, whereas the hotter nitrous oxide–acetylene flame (slot length, 50 mm) is reserved for the more 'refractory' elements, e.g. Al. It is common, therefore, to refer to this technique as flame atomic absorption spectroscopy or FAAS.

The introduction of an aqueous sample into the flame is achieved by using a pneumatic concentric nebulizer coupled with an expansion chamber. The pneumatic concentric nebulizer (Figure 2.9) consists of a stainless-steel tube through which a Pt/Ir capillary tube is located. The sample is drawn up through the capillary tube by the action of the oxidant gas (air) escaping through an exit

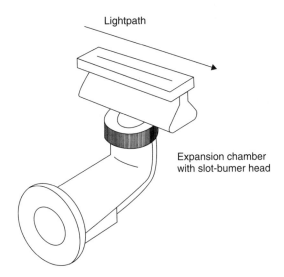

Figure 2.8 Schematic diagram of the slot-burner used in flame atomic absorption spectroscopy. From Dean, J. R., *Atomic Absorption and Plasma Spectroscopy*, 2nd Edition, ACOL Series, John Wiley & Sons, Limited, Chichester, UK, 1997. © University of Greenwich and reproduced by permission of the University of Greenwich.

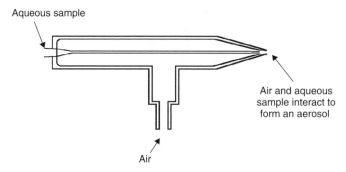

Aqueous sample

Air and aqueous
sample interact to
form an aerosol

Air

Figure 2.9 Schematic diagram of the pneumatic concentric nebulizer.

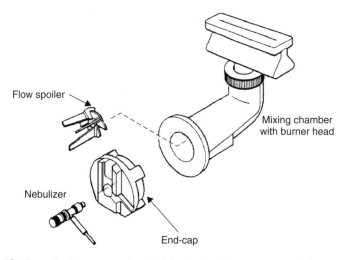

Flow spoiler

Mixing chamber
with burner head

Nebulizer

End-cap

Figure 2.10 The nebulizer–expansion (mixing) chamber system used for sample introduction in flame atomic absorption spectroscopy. From Dean, J. R., *Methods for Environmental Trace Analysis*, AnTS Series. Copyright 2003. © John Wiley & Sons, Limited. Reproduced with permission.

orifice located between the outside of the capillary tube and the inside of the stainless-steel tube. The interaction of escaping air with the liquid sample causes the formation of a coarse aerosol (this process is called the *Venturi effect*).

DQ 2.7

What function does the expansion chamber (Figure 2.10) fulfil?

Answer

The function of the expansion chamber is two-fold. The first is to convert the coarse aerosol generated by the nebulizer into a finer aerosol for transport to the slot-burner for atomization and allow residual aerosol particles to condense and go to waste. Secondly, the expansion chamber allows safe pre-mixing of the oxidant (air) and fuel (acetylene or nitrous oxide) gases prior to introduction into the slot-burner.

Another type of atomization cell is the graphite furnace and thus the technique which uses this cell is referred to as graphite furnace atomic absorption spectroscopy or GFAAS (*note*: sometimes this approach is also referred to as electrothermal atomic absorption spectroscopy or ETAAS). The graphite furnace is used when only a small amount of sample is available or when an increase in sensitivity is required. The graphite atomizer replaces the expansion chamber/slot-burner arrangement as the atomization cell. A small discrete sample ($5-100 \mu l$) is pipetted onto the inner surface of a graphite tube through a small opening (Figure 2.11). Light from the HCL passes through the graphite tube (typical dimensions are a length of $3-5$ cm with a diameter of $3-8$ mm). The passage of an electric current through the graphite tube, via water-cooled contacts, allows controlled heating. These heating stages (Figure 2.12) are pre-programmed into a controller to allow (a) solvent to be removed from the sample (drying), (b) removal of the sample matrix (ashing), (c) atomization of the element, (d) cleaning of the graphite tube of residual material and (e) cooling of the graphite tube in preparation for the next sample. It is the manner of these heating cycles (Table 2.2) that is the key to the success of the technique.

DQ 2.8

Why is the gas flow ON during the drying and ashing stages?

Answer
It is common practice for an internal inert gas (N_2 or Ar) to flow during the drying and ashing stages in order to remove any extraneous material.

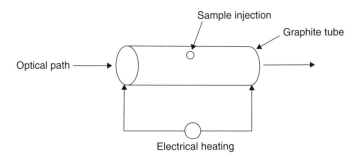

Figure 2.11 Schematic diagram of a typical graphite furnace used in atomic absorption spectroscopy. From Dean, J. R., *Atomic Absorption and Plasma Spectroscopy*, 2nd Edition, ACOL Series, John Wiley & Sons, Limited, Chichester, UK, 1997. © University of Greenwich and reproduced by permission of the University of Greenwich.

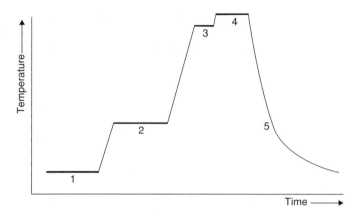

Figure 2.12 A typical time–temperature profile for graphite-furnace atomic absorption spectroscopy: 1, drying; 2, ashing; 3, atomization; 4, cleaning; 5, cooling. From Dean, J. R., *Atomic Absorption and Plasma Spectroscopy*, 2nd Edition, ACOL Series, John Wiley & Sons, Limited, Chichester, UK, 1997. © University of Greenwich and reproduced by permission of the University of Greenwich.

Table 2.2 Typical heating cycles in graphite-furnace atomic absorption spectroscopy

Stage	Temperature (°C)	Duration (s)	N_2 or Ar gas on or off
Drying	110	30	On
Ashing	400–1600	45	On
Atomization	2200	3	Off
Cleaning	2300	2	On
Cooling	—	60	On

DQ 2.9

Why is the gas flow OFF during the atomization stage?

Answer

It is common practice to have no flow of gas during the atomization stage in order to maximize the 'time-duration of atoms' within the graphite furnace.

DQ 2.10

Why might the ashing temperature vary?

Answer

In an ideal situation, organic matrix components are removed without any loss of the analyte of interest. However, this is not always possible and this is one of the major disadvantages of the technique. Remedies exist to preserve the analyte of interest by the addition of other chemicals/reagents which prevent premature loss of the analytes. This process is called *matrix modification*. For further information on this, consult Section 2.8.1.

DQ 2.11

During which stage is the absorbance (signal) recorded?

Answer

The absorption of the radiation source by the atomic vapour is measured during the atomization stage.

A specialized form of atomization cell is available for a limited number of elements that are capable of forming volatile hydrides. The technique which uses this type of cell is known as hydride-generation atomic absorption spectroscopy (HyAAS).

DQ 2.12

Which elements are capable of forming hydrides as used in HyAAS?

Answer

Arsenic (As), antimony (Sb), selenium (Se) and tin (Sn).

In HyAAS, an acidified solution of the sample is reacted with sodium tetraborohydride solution which liberates the gaseous hydride. Equation (2.3) illustrates the chemical generation of arsenic hydride (or arsine):

$$3BH_4^- + 3H^+ + 4H_3AsO_3 \longrightarrow 3H_3BO_3 + 4\mathbf{AsH_3} + 3H_2O \tag{2.3}$$
$$\text{(arsine)}$$

By using a gas–liquid separation device, the generated hydride is transported to an atomization cell (either an electrically heated or flame-heated quartz tube) by using a carrier gas. A variation of this approach, called *cold-vapour generation*, is exclusively reserved for the element mercury (i.e. cold-vapour atomic absorption spectroscopy, CVAAS). The mercury present in a sample is reduced, using tin(II) chloride, to elemental mercury (Equation (2.4)):

$$Sn^{2+} + Hg^{2+} \longrightarrow Sn^{4+} + \mathbf{Hg^0} \tag{2.4}$$
$$\text{(elemental mercury)}$$

This mercury vapour is then transported to an atomization cell (a long-pathlength glass absorption cell) by a carrier gas.

Wavelength separation is achieved by using a monochromator (Czerny–Turner configuration). This monochromator (Figure 2.13) has a focal length of 0.25–0.5 m with a grating containing only 600 lines mm^{-1} and a resolution of 0.2–0.02 nm. The attenuation of the HCL radiation by the atomic vapour is detected by a photomultiplier tube (PMT) (Figure 2.14). The PMT is a device for proportionally converting photons of light to electric current. Incident light

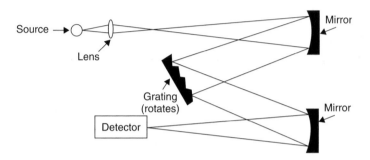

Figure 2.13 Schematic diagram of the optical layout of a spectrometer which incorporates the Czerny–Turner spectral mounting (configuration). From Dean, J. R., *Atomic Absorption and Plasma Spectroscopy*, 2nd Edition, ACOL Series, John Wiley & Sons, Limited, Chichester, UK, 1997. © University of Greenwich and reproduced by permission of the University of Greenwich.

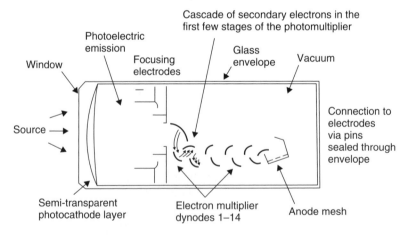

Figure 2.14 Schematic representation of the operation of a photomultiplier tube. From Dean, J. R., *Atomic Absorption and Plasma Spectroscopy*, 2nd Edition, ACOL Series, John Wiley & Sons, Limited, Chichester, UK, 1997. © University of Greenwich and reproduced by permission of the University of Greenwich.

strikes a photosensitive material which converts the light to electrons (this process is called the *photoelectric effect*). The generated electron is then focused and multiplied by a series of dynodes prior to collection at the anode. The multiplied electrons (or electrical current) are then converted into a voltage signal and then via an analogue-to-digital (A/D) converter into a suitable computer for processing.

The occurrence of molecular absorbance and scatter in AAS can be overcome by the use of background correction methods. Various types of background corrections are commonly used, including a continuum source, an HCL capable of changing between high and low electric currents (Smith–Hieftje background technique) and an approach based on the *Zeeman effect*.

SAQ 2.5

How does the Smith–Hieftje background correction method work?

Additionally, other problems which affect the recorded absorbance signal can occur and are based on chemical, ionization, physical and spectral interferences.

2.4 Atomic Emission Spectroscopy

The instrumentation used for atomic emission spectroscopy (AES) comprises an atomization cell, a spectrometer/detector and a read-out device. In its simplest form, i.e. the flame photometer (FP), the atomization cell consists of a flame (air–natural gas) while the spectrometer comprises an interference filter, followed by a photodiode or photoemissive detector.

DQ 2.13

What needs to be changed to measure potassium at 766.5 nm and sodium at 589.0 nm when using an FP?

Answer

The wavelength can be simply altered by changing the interference filter.

Modern instruments for AES, however, use an argon-based inductively coupled plasma (ICP) as the atomization cell. The technique is therefore referred to as inductively coupled plasma–atomic emission spectroscopy or ICP–AES. The ICP is formed within a plasma torch, i.e. three concentric glass tubes (Figure 2.15), known as the inner, intermediate (plasma) and external (coolant) tubes through which the argon gas passes. Argon gas enters the intermediate and external tubes via tangentially arranged entry points, while the sample aerosol is introduced in a carrier stream of argon, via the inner tube. Located around the external glass tube is a coil of copper tubing through which water is recirculated. Power input

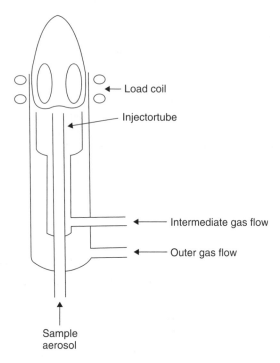

Figure 2.15 Schematic diagram of an inductively coupled plasma torch. From Dean, J. R., *Atomic Absorption and Plasma Spectroscopy*, 2nd Edition, ACOL Series, John Wiley & Sons, Limited, Chichester, UK, 1997. © University of Greenwich and reproduced by permission of the University of Greenwich.

to the ICP is achieved through this copper, load or induction coil, at a power of 1.3 kW (can range from 0.5–1.5 kW) at a frequency of either 27 or 40 MHz. The input power causes the induction of an oscillating magnetic field whose lines of force are axially orientated inside the plasma torch and follow elliptical paths outside the induction coil. To initiate the plasma, the carrier gas flow through the inner tube is switched off and a spark added momentarily from a Tesla coil, which is in contact with the outside of the plasma torch. This source of 'seed' electrons causes ionization of the argon gas. This process is 'self-sustaining', so that argon, argon ions and electrons co-exist (i.e. a 'plasma') within the confines of the plasma torch but protruding from its top in the shape of a bullet but emitting bright-white luminous light.

DQ 2.14

Why is it important never to look at the plasma light with the naked eye?

Answer

This is important as the plasma light causes damage due to the ultraviolet light that is emitted from the source.

The characteristic bullet-shape of the plasma is formed by the escaping high-velocity argon gas interacting with the surrounding atmosphere which causes air entrainment back towards the plasma torch itself. In order to introduce the sample aerosol into the confines of the hot plasma gas (7000–10 000 K), the inner-tube carrier gas is switched on and this 'punches a hole' into the centre of the plasma, thus creating the characteristic doughnut or toroidal shape of the ICP. The emitted radiation from the ICP can be viewed in a side-on (or lateral) or end-on (or axial) configuration (Figure 2.16).

Aqueous samples are introduced into an ICP–AES system by using a nebulizer/spray-chamber arrangement. The most common nebulizer is the pneumatic concentric glass nebulizer (Figure 2.17). The principle of operation of this nebulizer is identical to that described above for FAAS; the major differences are in the construction of the device and the gas used (for ICP–AES, the device is constructed entirely from glass to a higher specification). As the ICP is created from argon gas, so the gas of choice for use in this nebulizer is argon. As before, the interaction of aqueous sample and gas produces a coarse aerosol. Extensive research has been carried out over previous years to evaluate different nebulizers and so a range of devices are now available. A cross-flow nebulizer generates a coarse aerosol via two capillary needles positioned at 90° to each other with their tips not quite touching (Figure 2.18). Through one capillary tube, the argon carrier gas flows while through the other capillary the liquid sample is pumped. At the exit point, the force of the escaping argon gas is sufficient to 'shatter' the sample into a coarse aerosol. Another nebulizer (V-groove, Babington-type or Burgener nebulizer) allows aqueous samples with high-dissolved solids to be

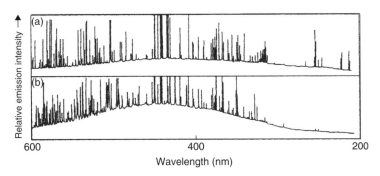

Figure 2.16 Comparison of the spectral features and background emission characteristics observed for (a) an axially viewed ICP and (b) a conventional, side-on viewed ICP. From Davies, R., Dean, J. R. and Snook, R. D., *Analyst*, **110**, 535–540 (1985). Reproduced by permission of The Royal Society of Chemistry.

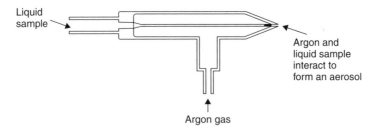

Liquid
sample

Argon and
liquid sample
interact to
form an aerosol

Argon gas

Figure 2.17 Schematic diagram of the pneumatic concentric glass nebulizer. From Dean, J. R., *Atomic Absorption and Plasma Spectroscopy*, 2nd Edition, ACOL Series, John Wiley & Sons, Limited, Chichester, UK, 1997. © University of Greenwich and reproduced by permission of the University of Greenwich.

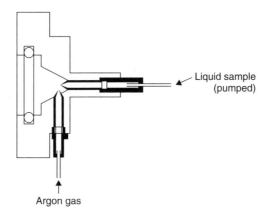

Liquid sample
(pumped)

Argon gas

Figure 2.18 Schematic diagram of the cross-flow nebulizer. From Dean, J. R., *Atomic Absorption and Plasma Spectroscopy*, 2nd Edition, ACOL Series, John Wiley & Sons, Limited, Chichester, UK, 1997. © University of Greenwich and reproduced by permission of the University of Greenwich.

analysed (Figure 2.19). An alternative approach is provided by the ultrasonic nebulizer which uses a vibrating piezoelectric transducer to transform the liquid sample into an aerosol.

The coarse aerosol generated by a nebulizer then requires to be converted into a finer aerosol. In ICP–AES, this is achieved by using a spray chamber.

DQ 2.15

What are the purposes of the spray chamber?

Answer

The purposes are to reduce the amount of aerosol reaching the plasma, decrease the turbulence associated with the nebulization process and reduce the aerosol particle size.

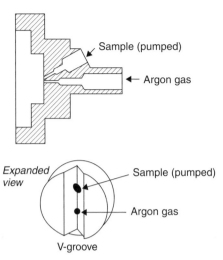

Figure 2.19 Schematic diagram of the V-groove, high-solids nebulizer. From Dean, J. R., *Atomic Absorption and Plasma Spectroscopy*, 2nd Edition, ACOL Series, John Wiley & Sons, Limited, Chichester, UK, 1997. © University of Greenwich and reproduced by permission of the University of Greenwich.

The most common spray-chamber designs are double-pass (Scott-type), cyclonic, single-pass and the direct (cylindrical) type. The most commonly used spray chamber is the double-pass. This consists of two concentric tubes, an inlet for the nebulizer, an exit for the finer aerosol and a waste drain (Figure 2.20). The latter is positioned to allow excess liquid (aerosol condensation) to flow to waste.

Alternative approaches for the direct introduction of sample into the plasma source are available and include electrothermal vaporization (ETV) and laser ablation (Figure 2.21). In both cases, the generated sample vapour is transported directly into the ICP torch, i.e. no nebulizer/spray chamber is required. Electrothermal vaporization is similar in operation to the graphite furnace used in AAS (described above).

DQ 2.16

What advantage might laser ablation offer?

Answer

Laser ablation allows the direct analysis of solid material. Laser light is directed onto the surface of a solid sample, so causing vaporization. The ablated material can then be transported direct, in a flow of argon, to the ICP torch.

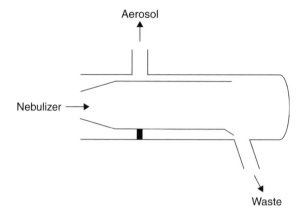

Figure 2.20 Schematic diagram of a double-pass spray chamber (Scott type). From Dean, J. R., *Atomic Absorption and Plasma Spectroscopy*, 2nd Edition, ACOL Series, John Wiley & Sons, Limited, Chichester, UK, 1997. © University of Greenwich and reproduced by permission of the University of Greenwich.

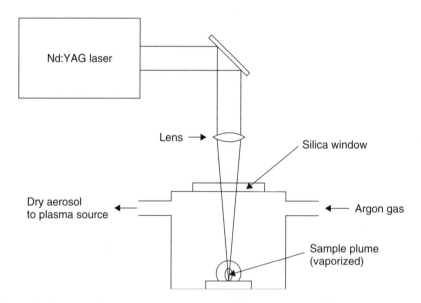

Figure 2.21 Schematic representation of a system used for laser ablation. From Dean, J. R., *Atomic Absorption and Plasma Spectroscopy*, 2nd Edition, ACOL Series, John Wiley & Sons, Limited, Chichester, UK, 1997. © University of Greenwich and reproduced by permission of the University of Greenwich.

One of the most common lasers used for this purpose is the Nd:YAG which operates in the near-infrared at 1064 nm.

SAQ 2.6

To what does Nd:YAG refer?

However, with optical frequency doubling, tripling, quadrupling and quintupling, the Nd:YAG laser can also be operated at the following wavelengths: 532, 355, 266 and 213 nm, respectively. In addition, excimer lasers have been used. As the operating wavelengths are dependent upon the gas, the choice of output wavelengths are 308, 248, 193 and 157 nm for XeCl, KrF, ArF and F_2, respectively.

In addition, hydride and cold-vapour techniques can be employed for the introduction of elements that form hydrides, plus mercury, directly into the ICP torch. The procedure is identical to that described for AAS (see Section 2.1.1).

Light emitted from the ICP source is focused onto the entrance slit of a spectrometer by using a convex-lens arrangement. The spectrometer separates the emitted light into its component wavelengths. In sequential analysis, one wavelength, corresponding to one element at a time, is measured.

DQ 2.17

To what does sequential multielement analysis refer?

Answer

Sequential multielement analysis allows several wavelengths to be measured at the same time.

The typical wavelength coverage of a spectrometer for ICP–AES is from 167 nm (Al) to 852 nm (Cs). Detection of the different wavelengths is achieved with either a photomultiplier tube (PMT) or charge-transfer device (CTD). A brief description of the PMT was provided in Section 2.1.1. A CTD is a semiconductor device consisting of a series of cells or pixels which accumulate charge when exposed to light. The amount of accumulated charge is then proportional to the amount of light that a particular pixel has been exposed to. CTDs exist in two forms, i.e. the charge-coupled device (CCD) and the charge-injection device (CID).

SAQ 2.7

In which forms do you commonly find CCD/CID devices?

Like all spectroscopic techniques, ICP–AES suffers from interferences, including 'spectral overlap'. The latter can be alleviated by either increasing the resolution of the spectrometer or by selecting an alternative spectral emission line.

2.5 Inorganic Mass Spectrometry

This discussion of inorganic mass spectrometry focuses on the inductively coupled plasma–mass spectrometry (ICP–MS) system only. A description of the ICP has been described previously (Section 2.4).

DQ 2.18

What was one of the key features that led to the development of ICP–MS?

Answer

The development of an efficient interface was key to the coupling of an ICP, operating at atmospheric pressure, with a mass spectrometer, operating under high vacuum (Figure 2.22).

The interface consists of a water-cooled outer nickel sampling cone positioned in close proximity to the plasma source (Figure 2.23). The region behind the

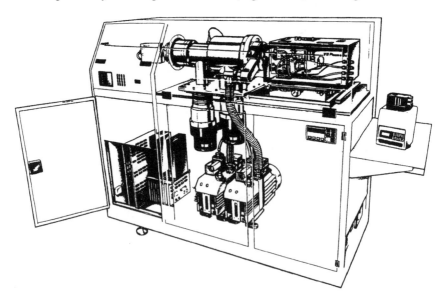

Figure 2.22 A commercially available inductively coupled plasma–mass spectrometer. From 'Applications Literature' published by VG Elemental. Reproduced by permission of ThermoFisher Scientific, Macclesfield, Cheshire, UK.

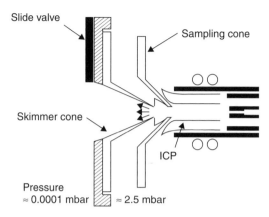

Figure 2.23 Schematic diagram of the inductively coupled plasma–mass spectrometer interface. From Dean, J. R., *Atomic Absorption and Plasma Spectroscopy*, 2nd Edition, ACOL Series, John Wiley & Sons, Limited, Chichester, UK, 1997. © University of Greenwich and reproduced by permission of the University of Greenwich.

sampling cone is maintained at a moderate pressure (approximately 2.5 mbar) by using a rotary vacuum pump.

SAQ 2.8

Can you express 2.5 mbar in other units of pressure?

As the gas flow through the sample cone is large, a second (skimmer) cone is placed close enough behind the sampling cone to allow the central portion of the expanding jet of plasma gas and ions to pass through the skimmer cone. The pressure behind the skimmer cone is maintained at approximately 10^{-4} mbar. The extracted ions are then focused by a series of electrostatic lenses into the mass spectrometer. If the ICP–MS system contains a collision/reaction cell (see Section 2.5.1), it is located before the mass spectrometer.

The mass spectrometer acts as a filter, transmitting ions with a pre-selected mass/charge ratio which are then detected and displayed. A range of mass spectrometers are available, including the quadrupole, high-resolution, ion-trap and time-of-flight spectrometers.

The quadrupole analyser consists of four straight metal rods positioned parallel to and equidistant from the central axis (Figure 2.24). By applying DC and RF voltages to opposite pairs of the rods, it is possible to have a situation where the DC voltage is positive for one pair and negative for the other. Likewise, the RF voltages on each pair are $180°$ out of phase, i.e. they are opposite in sign, but with the same amplitude. Ions entering the quadrupole are subjected to oscillatory paths by the RF voltage. However, by selecting appropriate RF and DC voltages,

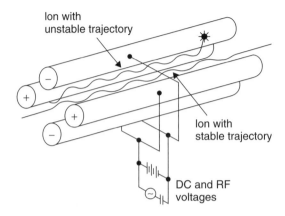

Figure 2.24 Schematic arrangement of the quadrupole analyser arrangement. From Dean, J. R., *Atomic Absorption and Plasma Spectroscopy*, 2nd Edition, ACOL Series, John Wiley & Sons, Limited, Chichester, UK, 1997. © University of Greenwich and reproduced by permission of the University of Greenwich.

only ions of a given mass/charge ratio will be able to traverse the length of the rods and emerge at the other end.

DQ 2.19

What happens to the other ions?

Answer

These ions are lost within the quadrupole analyser – as their oscillatory paths are too large, they collide with the rods and become neutralized.

A high-resolution mass spectrometer can be used to separate interferences. It consists of an electrostatic analyser (ESA) and a magnetic analyser (MA). The ESA consists of two curved plates to which is applied a DC voltage; the inner plate has a negative polarity which attracts ions, while the outer plate has a positive polarity which repels the ions. Therefore, only ions with a specific energy can pass through the ESA. The application of a particular field strength to the MA allows only ions with a specific m/z ratio to be separated.

An ion-trap mass spectrometer consists of a cylindrical ring electrode with two end-cap electrodes (Figure 2.25). The upper end-cap allows ions to be introduced into the trap where they are retained ('trapped'). The application of an RF voltage to the ring electrode stabilizes and retains ions of different m/z ratios. By then

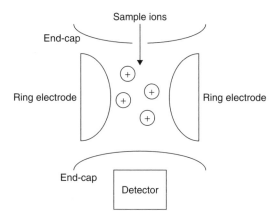

Figure 2.25 Schematic representation of the operation of an ion-trap mass spectrometer. From Dean, J. R., *Practical Inductively Coupled Plasma Spectroscopy*, AnTS Series. Copyright 2005. © John Wiley & Sons, Limited. Reproduced with permission.

increasing the applied voltage, the paths of ions of successive m/z ratios are made unstable and exit the trap via the lower end-cap.

A time-of-flight (TOF) mass spectrometer separates ions with respect to their different velocities (Figure 2.26). This is achieved by applying an accelerating voltage to the ions, hence resulting in them having different velocities. Lighter ions, which are travelling faster, reach the detector before the heavier ions. To increase the pathlength, a 'reflectron' is introduced into the mass analyser, which doubles the pathlength by reversing the direction of flow of the ions. It is common to find a TOF mass spectrometer used in conjunction with a quadrupole mass analyser.

Mass spectrometers separate ions – in order to measure a signal, this requires a method of detection. The most common type of detector is the continuous-dynode electron multiplier (Figure 2.27). The electron multiplier is 'horn-shaped' with a semiconductor (lead oxide)-coated internal surface. Detection is achieved by applying a high negative potential ($-3\,\text{kV}$) at the entrance of the horn, while the collector is held at ground potential. Incoming ions are attracted towards the negative potential of the horn and captured. The impact of the ions on the internal coated surface causes secondary electrons to be released. As the secondary electrons are produced, they are attracted towards the grounded collector. This results in these secondary electrons interacting further with the internal surface coating, so producing further electrons. Thus, electrons are 'multiplied' prior to collection where they are further amplified exterior to the device and recorded as a signal ('counts per second'). Other detectors used in mass spectrometers are the Faraday cup and the Daly detector.

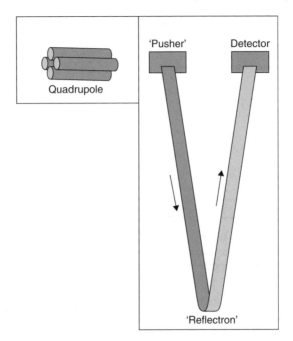

Figure 2.26 Schematic diagram of the layout of a time-of-flight mass spectrometer. From Dean, J. R., *Practical Inductively Coupled Plasma Spectroscopy*, AnTS Series. Copyright 2005. © John Wiley & Sons, Limited. Reproduced with permission.

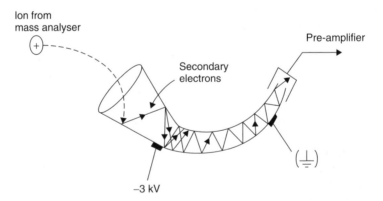

Figure 2.27 Schematic representation of the operating principles of the electron multiplier tube arrangement. From Dean, J. R., *Atomic Absorption and Plasma Spectroscopy*, 2nd Edition, ACOL Series, John Wiley & Sons, Limited, Chichester, UK, 1997. © University of Greenwich and reproduced by permission of the University of Greenwich.

2.5.1 Interferences in ICP–MS

Spectral interferences in ICP–MS result from isobaric (overlap of atomic masses of different elements) and molecular (polyatomic and doubly charged) species.
  Isobaric interferences are well characterized (Table 2.3).

DQ 2.20

How can isobaric interferences be avoided?

Answer

These interferences can be avoided for most elements by selecting an alternative isotope.

Table 2.3 Isobaric interferences from Period 4 of the Periodic Table. From Dean, J. R., *Atomic Absorption and Plasma Spectroscopy*, 2nd Edition, ACOL Series, John Wiley & Sons, Limited, Chichester, UK, 1997. © University of Greenwich and reproduced by permission of the University of Greenwich

Atomic mass	Element of interest (% abundance)	Interfering element (% abundance)
39	K (93.10)	—
40	Ca (96.97)	Ar (99.6)[a]; K (0.01)
41	K (6.88)	—
42	Ca (0.64)	—
43	Ca (0.14)	—
44	Ca (2.06)	—
45	Sc (100)	—
46	—	Ca (0.003); Ti (7.93)
47	—	Ti (7.28)
48	Ti (73.94)	Ca (0.19)
49	—	Ti (5.51)
50	—	Ti (5.34); V (0.24); Cr (4.31)
51	V (99.76)	—
52	Cr (83.76)	—
53	—	Cr (9.55)
54	—	Cr (2.38); Fe (5.82)
55	Mn (100)	—
56	Fe (91.66)	—
57	—	Fe (2.19)
58	Ni (67.88)	Fe (0.33)
59	Co (100)	—
60	Ni (26.23)	—
61	—	Ni (1.19)
62	—	Ni (3.66)
63	Cu (69.09)	—

(*continued overleaf*)

Table 2.3 (*continued*)

Atomic mass	Element of interest (% abundance)	Interfering element (% abundance)
64	Zn (48.89)	Ni (1.08)
65	Cu (30.91)	—
66	Zn (27.81)	—
67	—	Zn (4.11)
68	—	Zn (18.57)
69	Ga (60.40)	—
70	—	Zn (0.62); Ge (20.52)
71	Ga (39.60)	—
72	Ge (27.43)	—
73	—	Ge (7.76)
74	Ge (36.54)	Se (0.87)
75	As (100)	—
76	—	Ge (7.76); Se (9.02)
77	—	Se (7.58)
78	Se (23.52)	Kr (0.35)
79	Br (50.54)	—
80	Se (49.82)	Kr (2.27)
81	Br (49.46)	—
82	—	Se (9.19); Kr (11.56)
83	Kr (11.55)	—
84	Kr (56.90)	Sr (0.56)[b]
85	—	—
86	—	Kr (17.37); Sr (9.86)[b]

[a] Not in Period 4 of the Periodic Table but included because of its origin from the plasma source.
[b] Not in Period 4 of the Periodic Table but included for completeness.

For those elements without an alternative isotope, the possibilities are limited, for example, in the determination of calcium by ICP–MS. Both Ca and Ar (plasma gas) have major isotopes of atomic mass 40 (Ar 99.6% abundant, while Ca is 97.0% abundant). Two options are possible, i.e. select another mass for Ca (Ca is 2.1% abundant with an atomic mass unit (amu) of 44), or use a collision/reaction cell to either remove the interference or move Ca to another atomic mass (see below).

Polyatomic interferences (Table 2.4) are derived as a result of interactions between elements and the sample solution or plasma gas (Ar). As a result, it is possible to identify species, such as $^{35}Cl^{16}O^{+}$ at mass 51 (mass for the determination of vanadium) and $^{40}ArI^{16}O^{+}$ at mass 56 (mass for the determination of iron). In addition, doubly charged interferences can result. To understand this type of interference, it is worth remembering that a mass spectrometer separates on the basis of an element's mass/charge or m/z ratio. For example, barium (Ba) has atomic masses of 130, 132, 134, 135, 136, 137 and 138; however, Ba is capable of forming doubly charged species.

Table 2.4 Potential polyatomic interferences derived from the element of interest and its associated aqueous solution, the plasma gas itself and the type of acid(s) used to digest or prepare the sample. From Dean, J. R., *Atomic Absorption and Plasma Spectroscopy*, 2nd Edition, ACOL Series, John Wiley & Sons, Ltd, Chichester, UK, 1997. © University of Greenwich and reproduced by permission of the University of Greenwich

Atomic mass	Element of interest (% abundance)	Polyatomic interference
39	K (93.10)	$^{38}Ar^1H^+$
40	Ca (96.97)	$^{40}Ar^+$
41	K (6.88)	$^{40}Ar^1H^+$
42	Ca (0.64)	$^{40}Ar^2H^+$
43	Ca (0.14)	—
44	Ca (2.06)	$^{12}C^{16}O^{16}O^+$
45	Sc (100)	$^{12}C^{16}O^{16}O^1H^+$
46	—	$^{14}N^{16}O^{16}O^+$; $^{32}S^{14}N^+$
47	—	$^{31}P^{16}O^+$; $^{33}S^{14}N^+$
48	Ti (73.94)	$^{31}P^{16}O^1H^+$; $^{32}S^{16}O^+$; $^{34}S^{14}N^+$
49	—	$^{32}S^{16}O^1H^+$; $^{33}S^{16}O^+$; $^{14}N^{35}Cl^+$
50	—	$^{34}S^{16}O^+$; $^{36}Ar^{14}N^+$
51	V (99.76)	$^{35}Cl^{16}O^+$; $^{34}S^{16}O^1H^+$; $^{14}N^{37}Cl^+$; $^{35}Cl^{16}O^+$
52	Cr (83.76)	$^{40}Ar^{12}C^+$; $^{36}Ar^{16}O^+$; $^{36}S^{16}O^+$; $^{35}Cl^{16}O^1H^+$
53	—	$^{37}Cl^{16}O^+$
54	—	$^{40}Ar^{14}N^+$; $^{37}Cl^{16}O^1H^+$
55	Mn (100)	$^{40}Ar^{14}N^1H^+$
56	Fe (91.66)	$^{40}Ar^{16}O^+$
57	—	$^{40}Ar^{16}O^1H^+$
58	Ni (67.88)	—
59	Co (100)	—
60	Ni (26.23)	—
61	—	—
62	—	—
63	Cu (69.09)	$^{31}P^{16}O_2^+$
64	Zn (48.89)	$^{31}P^{16}O_2^1H^+$; $^{32}S^{16}O^{16}O^+$; $^{32}S^{32}S^+$
65	Cu (30.91)	$^{33}S^{16}O^{16}O^+$; $^{32}S^{33}S^+$
66	Zn (27.81)	$^{34}S^{16}O^{16}O^+$; $^{32}S^{34}S^+$
67	—	$^{35}Cl^{16}O^{16}O^+$
68	—	$^{40}Ar^{14}N^{14}N^+$; $^{36}S^{16}O^{16}O^+$; $^{32}S^{36}S^+$
69	Ga (60.40)	$^{37}Cl^{16}O^{16}O^+$
70	—	$^{35}Cl_2^+$; $^{40}Ar^{14}N^{16}O^+$
71	Ga (39.60)	$^{40}Ar^{31}P^+$; $^{36}Ar^{35}Cl^+$
72	Ge (27.43)	$^{37}Cl^{35}Cl^+$; $^{36}Ar^{36}Ar^+$; $^{40}Ar^{32}S^+$
73	—	$^{40}Ar^{33}S^+$; $^{36}Ar^{37}Cl^+$

(*continued overleaf*)

Table 2.4 (*continued*)

Atomic mass	Element of interest (% abundance)	Polyatomic interference
74	Ge (36.54)	$^{37}Cl^{37}Cl^+$; $^{36}Ar^{38}Ar^+$; $^{40}Ar^{34}S^+$
75	As (100)	$^{40}Ar^{35}Cl^+$
76	—	$^{40}Ar^{36}Ar^+$; $^{40}Ar^{36}S^+$
77	—	$^{40}Ar^{37}Cl^+$; $^{36}Ar^{40}Ar^1H^+$
78	Se (23.52)	$^{40}Ar^{38}Ar^+$
79	Br (50.54)	$^{40}Ar^{38}Ar^1H^+$
80	Se (49.82)	$^{40}Ar^{40}Ar^+$
81	Br (49.46)	$^{40}Ar^{40}Ar^1H^+$
82	—	$^{40}Ar^{40}Ar^1H^1H^+$
83	Kr (11.55)	—
84	Kr (56.90)	—
85	—	—
86	—	—

DQ 2.21

At what m/z ratio will the resultant doubly charged species for $^{138}Ba^+$ occur?

Answer

The resultant doubly charged species for $^{138}Ba^+$ will occur at atomic mass 69, i.e. Ba^{2+} (m/z, where z (the charge) is 2).

2.5.1.2 Collision and Reaction Cells

Collision/reaction cells are located behind the sample/skimmer cone arrangement, but *before* the mass analyser. Collision and reaction cells allow for interferences or element ion m/z ratio shifts. A range of processes can be used, including 'charge-exchange'. In the latter, the argon plasma gas ion interference can be removed with the formation of uncharged argon plasma gas, which cannot then be detected:

$$Ca^+ + Ar^+ + H_2 \longrightarrow Ca^+ + Ar + H_2^+ \qquad (2.5)$$

The use of collision/reaction cells in ICP–MS can result in the 'chemical resolution' of isobaric and polyatomic interferences.

SAQ 2.9

How might 'chemical resolution' be applied in a collision/reaction cell?

2.6 X-Ray Fluorescence Spectroscopy

X-ray fluorescence (XRF) spectroscopy is used to measure elements in solid and aqueous samples, either qualitatively or quantitatively. It main advantage compared to the techniques already discussed is its ability to determine elements in solid samples with minimal sample preparation. The principle of the technique is based on the 'relocation' of electrons within the atomic structure of the elements under investigation. If an atom within a sample is irradiated with X-rays, an electron from its inner shell is lost (Figure 2.28(a)). In order for the atom to resume its most stable energy configuration this requires an electron from a higher-energy level to fill the vacancy (Figure 2.28b). This transfer of an electron from a higher-energy state to a lower-energy state results in the emission of an X-ray characteristic of that element. Two different instruments for XRF spectroscopy are available, dependent upon how the radiation is monitored. In

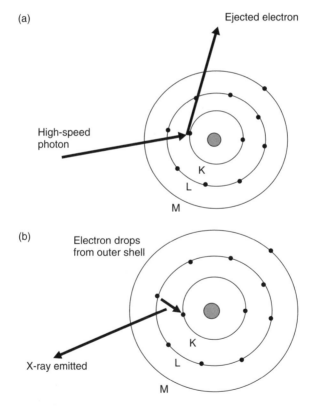

Figure 2.28 Principles of X-ray fluorescence spectroscopy: (a) loss of an electron from the inner shell of an atom of an element after irradiation with X-rays; (b) transfer of an electron from a higher-energy state to fill the vacancy created (lower-energy state), resulting in the emission of an X-ray characteristic of that element.

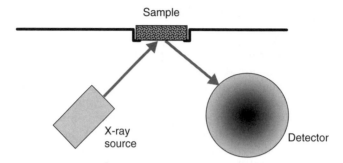

Figure 2.29 Schematic representation of the layout of an energy-dispersive X-ray fluorescence spectrometer.

wavelength-dispersive XRF (WDXRF), the emitted radiation from the element is dispersed by a crystal into its component wavelengths. Sequential scanning of the different wavelengths allows identification of different elements within the sample. Alternatively, in energy-dispersive XRF (EDXRF), all of the emitted radiation is measured at a detector and identified electronically.

An EDXRF spectrometer (Figure 2.29) requires a source of X-rays (to excite the atoms in the sample), a sample holder and a spectrometer to measure the energy (or wavelength) and intensity of the radiation emitted by each element in the sample.

Liquid samples are placed directly in a small holder (2–4 cm in diameter), covered on one side with a sample support film, e.g. Mylar™.

DQ 2.22

Why do you think the use of a special support film is required?

Answer

The use of such a film enables the liquid sample to remain within the sample holder and provides a transparent 'window' for the X-rays to pass through.

Solid objects can be placed directly into the sample chamber of the XRF spectrometer, provided that they are smooth and flat. Similarly, loose powder samples can be placed in the sample holder and analysed directly. However, a better approach for loose powder samples is to mix them with a binder and press into a pellet prior to analysis. The sample is then placed inside the chamber of the spectrometer. While the analysis can be carried out in air (air absorbs

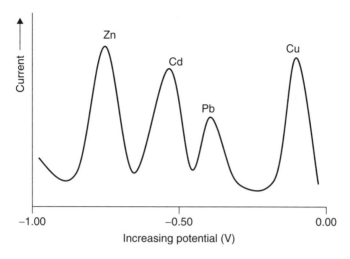

Figure 2.30 An example of a typical anodic stripping voltammogram.

low-energy X-rays from elements below atomic number 20, e.g. calcium), a vacuum or helium purge is often used for improved performance. X-rays are then directed at the sample and the resultant emission is detected by a Si(Li) detector.

2.7 Electrochemistry

Anodic stripping voltammetry (ASV) is an electrochemical technique for the analysis of trace metals in solution. Three electrodes (working electrode, reference electrode and a counter electrode) are positioned in an electrolytic cell. The sample is placed in the cell along with 0.1 mol l^{-1} acetate buffer at pH 4.5 (supporting electrolyte). As the signal is affected by the dissolved oxygen, this is removed from solution by bubbling an inert gas (e.g. nitrogen or argon) through the cell. By holding the working electrode (e.g. a mercury-drop electrode) at a small negative potential (with respect to the reference electrode), any metal ions in solution are attracted to the electrode and deposited. The process can be aided by stirring of the solution. After a suitable (deposition) time, the potential of the working electrode is scanned in the positive direction (i.e. to a lower potential). This process allows deposited metals on the surface of the working electrode to become oxidized and re-introduced into solution. Plotting the change in current between the working electrode and the counter electrode allows a signal to be recorded (Figure 2.30).

2.8 Hyphenated Techniques

The term 'hyphenated technique' is often used to mean the linking of one analytical instrument with another. In this context, it is not unreasonable to discuss
the linking of a gas chromatograph or a high performance liquid chromatograph
with a mass spectrometer (as described in Chapter 3, Section 3.3). However, this
section seeks to identify a hyphenated technique in the context of the linking of
various atomic spectroscopic techniques with chromatography, in particular for
the acquisition of elemental species-related information, e.g. the oxidation state
of the element, detection of organometallic compounds or the identification of
elements that are incorporated in protein structures.

SAQ 2.10

Why might the identification of element-species information be important?

The interfacing of an high performance liquid chromatograph with an ICP
normally offers few challenges with respect to physical coupling of the two
techniques.

DQ 2.23

Why do you think the coupling of an HPLC instrument to an ICP is
'easy'?

Answer

This is because the HPLC flow rates are typically in the range of 1 ml
min^{-1}, which is compatible with the aspiration rates of a nebulizer/spray
chamber used for ICP analysis. No instrumental modifications are therefore required to connect an HPLC system to an ICP system, apart from
a short length of polytetrafluoroethylene (PTFE) or poly(ether ether
ketone) (PEEK) tubing.

However, it is worth remembering that the nebulizer/spray chamber is an inef-
ficient interface for the ICP, resulting in < 2% of the sample being available for
excitation and/or ionization. The presence of mobile phase as the carrier for the
sample can also lead to potential interferences when either AES or MS is used
for detection (see also Sections 2.2 and 2.3.1). For example, the presence of a
buffer in the mobile phase can lead to isobaric interferences in ICP–MS, while
the presence of organic solvent can lead to molecular species, e.g. C_2, interfering
with analysis in ICP–AES.

The interfacing of a gas chromatograph with an ICP is relatively straightforward, although the issues are different. As GC separates volatile compounds in

a carrier-gas stream, the key aspect of the interface with an ICP is to retain the separated compounds in the gaseous form, *enroute* to the plasma.

DQ 2.24

How do you think the compounds can be retained in gaseous form?

Answer

This is achieved by using a heated transfer line (200–250°C), which allows the carrier gas (argon, helium or nitrogen) and volatile compounds to be directly transported to the ICP.

In GC–ICP analysis, no nebulizer/spray chamber is required, as the end of the heated transfer line is inserted directly into the injector tube of the ICP torch. This obviously results in a high-transport efficiency compared to HPLC–ICP and correspondingly an enhanced sensitivity for the elements analysed in the separated compounds.

2.9 Comparison of Elemental Analytical Techniques

Comparisons of the various elemental analytical techniques by performing SWOT (strengths, weaknesses, opportunities and threats) analyses are shown in Tables 2.5–2.10.

Table 2.5 SWOT analysis: elemental analysis (in general terms)

Strengths	• Range of elements can be determined at major, minor, trace and 'ultra-trace' levels
	• Techniques applicable to a wide range of sample types and applications (industrial, chemical, pharmaceutical, clinical and forensic)
Weaknesses	• Calibration strategies always required
	• Some form of sample preparation is always required. Sample preparation techniques vary depending upon sample type
	• Interferences often need to be overcome/circumvented
	• High-purity materials need to be used to prepare calibration standards
	• Linearity of techniques needs to be established
Opportunities	• Coupling of chromatography to elemental analysis techniques leads to enhanced information, e.g. chemical speciation data
	• Techniques available in a range of formats, providing a diverse range of opportunities
Threats	• Long-term threat from miniaturization
	• High capital/running costs of (some) instruments

Table 2.6 SWOT analysis: atomic absorption spectroscopy (AAS)

Strengths	• Available in a range of atomization cells (flame, graphite-furnace, cold-vapour and hydride-generation) which allow major, minor and trace elements to be determined • Relatively low cost compared to other techniques • Small samples volumes for GFAAS
Weaknesses	• Only possible to determine one element at a time. Different hollow-cathode lamp required per element • Samples need to be in aqueous form • Interferences: spectral, chemical and ionization, but well-identified and characterized • Sample pre-concentration may be required for certain applications • Different flame composition required for more refractory elements (FAAS) • Relatively large sample volumes required for FAAS • Matrix modifiers required for GFAAS
Opportunities	• High sensitivity for difficult elements, e.g. cold-vapour generation for mercury, hydride-generation for arsenic, selenium, etc. • A range of atomization cell formats is available which allow individual instruments to be dedicated to particular applications
Threats	• Single element analysis compared to other techniques • Too many accessories, i.e. use of different atomization cells and hollow-cathode lamp changes, required in a commercial laboratory

Table 2.7 SWOT analysis: inductively coupled plasma–atomic emission spectroscopy (ICP–AES)

Strengths	• Simultaneous multielement analysis at major, minor and trace levels • Ease of coupling alternative formats, e.g. laser ablation, flow-injection, hydride-generation and chromatography • Large linear dynamic range
Weaknesses	• Moderate–high cost to purchase instrumentation • Significant operating costs (argon) • Samples need to be in aqueous form (except for laser ablation) • Interferences – spectral. Complex spectra obtained for certain elements, e.g. rare-earth elements
Opportunities	• Ability to provide trace element chemical speciation information by direct coupling with GC or HPLC
Threats	• Major threat from ICP–MS

Table 2.8 SWOT analysis: inductively coupled plasma–mass spectrometry (ICP–MS)

Strengths	• Simultaneous multielement analysis at major, minor, trace and 'ultra-trace' levels
	• Ease of coupling alternative formats, e.g. laser ablation, flow-injection, hydride-generation and chromatography
	• Relatively simple spectra for most elements
	• Possible to use stable isotopes for tracer studies; isotope dilution analysis for quantitative analysis
	• Use of collision/reaction cells to remove molecular interferences
	• Large linear dynamic range
Weaknesses	• High cost to purchase instrumentation
	• Significant operating costs (argon)
	• Some elements suffer from isobaric and/or molecular interferences
	• Samples need to be in aqueous form (except for laser ablation)
Opportunities	• Ability to provide (ultra)-trace element chemical speciation information by direct coupling with GC or HPLC
Threats	• Higher skill level required by operator

Table 2.9 SWOT analysis: X-ray fluorescence (XRF) spectroscopy

Strengths	• Multielement analysis at major, minor and trace levels
	• Analysis can be done directly on (semi)-solid or aqueous samples
	• Minimal sample preparation required
Weaknesses	• Interpretation of data/fundamentals requires advanced knowledge
	• Cannot easily increase the sensitivity
Opportunities	• Direct analysis of (semi)-solid samples
Threats	• Use of X-rays!

Table 2.10 SWOT analysis: anodic stripping voltammetry (ASV)

Strengths	• Sequential multielement analysis at major, minor and trace levels
	• Small sample size
Weaknesses	• Limited to a few elements
	• Often requires method of standard additions to obtain accurate data
	• Samples must be in aqueous form
Opportunities	• Low cost of instrumentation
Threats	• Considered to be a difficult and unreliable technique (electroanalytical)

2.10 Selected Resources on Elemental Analytical Techniques

2.10.1 Specific Books on Atomic Spectroscopy[†]

Dean, J. R., *Practical Inductively Coupled Plasma Spectroscopy*, AnTS Series, John Wiley and Sons, Ltd, Chichester, UK, 2005.

Lajunen, L. H. J. and Peramaki, P., *Spectrochemical Analysis by Atomic Absorption and Emission*, 2nd Edition, The Royal Society of Chemistry, Cambridge, UK, 2004.

Holland, J. G. and Tanner, S. D. (Eds), *Plasma Source Mass Spectrometry: Applications and Emerging Technologies*, The Royal Society of Chemistry, Cambridge, UK, 2003.

Nolte, J., *ICP Emission Spectrometry: A Practical Guide*, Wiley-VCH, Weinheim, Germany, 2003.

Thomas, R., *Practical Guide to ICP–MS*, Marcel Dekker, New York, NY, USA, 2003.

Broekaert, J., *Analytical Atomic Spectrometry with Flames and Plasmas*, Wiley-Interscience, New York, NY, USA, 2002.

Beauchemin, D., Gregoire, D. C., Karanassios, V., Wood, T. J. and Mermet, J. M., *Discrete Sample Introduction Techniques for Inductively Coupled Plasma–Mass Spectrometry*, Elsevier, Amsterdam, The Netherlands, 2000.

Caruso, J. A., Sutton, K. L. and Ackley, K. L., *Elemental Speciation: New Approaches for Trace Element Analysis*, Elsevier, Amsterdam, The Netherlands, 2000.

Taylor, H., *Inductively Coupled Plasma–Mass Spectrometry: Practices and Techniques*, Academic Press, London, UK, 2000.

Ebdon, L., Evans, E. H., Fisher, A. S. and Hill, S. J., *An Introduction to Analytical Atomic Spectrometry*, John Wiley & Sons, Ltd, Chichester, UK, 1998.

Montaser, A., *Inductively Coupled Plasma–Mass Spectrometry*, 2nd Edition, Wiley-VCH, Weinheim, Germany, 1998.

Dean, J. R., *Atomic Absorption and Plasma Spectroscopy*, 2nd Edition, ACOL Series, John Wiley & Sons, Ltd, Chichester, UK, 1997.

Lobinski, R. and Marczenko, Z., *Spectrochemical Trace Analysis for Metals and Metalloids*, in *Wilson and Wilsons Comprehensive Analytical Chemistry*, Vol. XXX, Weber, S. G. (Ed.), Elsevier, Amsterdam, The Netherlands, 1996.

[†] Arranged in chronological order.

Cresser, M. S., *Flame Spectrometry in Environmental Chemical Analysis: A Practical Approach*, The Royal Society of Chemistry, Cambridge, UK, 1995.

Dedina, J. and Tsalev, D. I., *Hydride Generation Atomic Absorption Spectrometry*, John Wiley & Sons, Ltd, Chichester, UK, 1995.

Evans, E. H., Giglio, J. J., Castillano, T. M. and Caruso, J. A., *Inductively Coupled and Microwave Induced Plasma Sources for Mass Spectrometry*, The Royal Society of Chemistry, Cambridge, UK, 1995.

Fang, Z., *Flow Injection Atomic Absorption Spectrometry*, John Wiley & Sons, Ltd, Chichester, UK, 1995.

Howard, A. G. and Statham, P. J., *Inorganic Trace Analysis. Philosophy and Practice*, John Wiley & Sons, Ltd, Chichester, UK, 1993.

Slickers, K., *Automatic Atomic Emission Spectroscopy*, 2nd Edition, Bruhlsche Universitatsdruckerei, Giessen, Germany, 1993.

Vandecasteele. C. and Block, C. B., *Modern Methods of Trace Element Determination*, John Wiley & Sons, Ltd, Chichester, UK, 1993.

Jarvis, K. E., Gray, A. L. and Houk, R. S., *Handbook of Inductively Coupled Plasma–Mass Spectrometry*, Blackie and Son, Glasgow, Scotland, UK, 1992.

Lajunen, L. H. J., *Spectrochemical Analysis by Atomic Absorption and Emission*, The Royal Society of Chemistry, Cambridge, UK, 1992.

Holland, G. and Eaton, A. N., *Applications of Plasma Source Mass Spectrometry*, The Royal Society of Chemistry, Cambridge, UK, 1991.

Jarvis, K. E., Gray, A. L., Jarvis, I. and Williams, J. G., *Plasma Source Mass Spectrometry*, The Royal Society of Chemistry, Cambridge, UK, 1990.

Sneddon, J., *Sample Introduction in Atomic Spectroscopy*, Vol. 4, Elsevier, Amsterdam, The Netherlands, 1990.

Date, A. R. and Gray, A. L., *Applications of Inductively Coupled Plasma–Mass Spectrometry*, Blackie and Son, Glasgow, Scotland, UK, 1989.

Harrison, R. M. and Rapsomanikis, S., *Environmental Analysis Using Chromatography Interfaced with Atomic Spectroscopy*, Ellis Horwood Ltd, Chichester, UK, 1989.

Moenke-Blankenburg, L., *Laser Microanalysis*, John Wiley & Sons, Inc., New York, NY, USA, 1989.

Moore, G. L., *Introduction to Inductively Coupled Plasma–Atomic Emission Spectroscopy*, Elsevier, Amsterdam, The Netherlands, 1989.

Thompson, M. and Walsh, J. N., *A Handbook of Inductively Coupled Plasma Spectrometry*, 2nd Edition, Blackie and Son, Glasgow, Scotland, UK, 1989.

Adams, F., Gijbels, R. and van Grieken, R., *Inorganic Mass Spectrometry*, John Wiley & Sons, Inc., New York, NY, USA, 1988.

Ingle, J. D. and Crouch, S. R., *Spectrochemical Analysis*, Prentice-Hall International, London, UK, 1988.

Boumans P. W. J. M., *Inductively Coupled Plasma–Emission Spectrometry*, Parts 1 and 2, John Wiley & Sons, Inc., New York, NY, USA, 1987.

Montaser, A. and Golightly, D. W., *Inductively Coupled Plasmas in Analytical Atomic Spectrometry*, VCH Publishers, New York, NY, USA, 1987.

Welz, B., *Atomic Absorption Spectrometry*, VCH, Weinheim, Germany, 1985.

Ebdon, L., *An Introduction to Atomic Absorption Spectroscopy*, Heyden and Son, London, UK, 1982.

2.10.2 Specific Books on Electroanalytical Techniques

Wang, J., *Analytical Electrochemistry*, 2nd Edition, John Wiley & Sons, Ltd, Chichester, UK, 2000.

Summary

The different sample preparation techniques applied to solid and liquid samples for elemental analysis are described. In addition, the various analytical techniques capable of performing quantitative analysis of metals/metalloids in environmental samples are reviewed. Finally, SWOT (strengths, weaknesses, opportunities and threats) analyses are described in order to highlight the parameters that should be considered when choosing between the different analytical techniques.

Chapter 3

Sample Preparation and Analytical Techniques for Persistent Organic Pollutant Analysis of Environmental Contaminants

Learning Objectives

- To appreciate the different approaches available for preparation of solid and liquid samples for persistent organic pollutant analysis.
- To be aware of the procedure for preparing a solid sample using a range of options, including Soxhlet, shake-flask, ultrasonic, supercritical fluid extraction, microwave-assisted extraction and pressurized fluid extraction.
- To be aware of the procedures for preparing liquid samples, including solvent extraction, solid-phase extraction and solid-phase microextraction.
- To understand and explain the principles of operation of a gas chromatography system.
- To be aware of the range of detectors that can be used for gas chromatography.
- To understand and explain the principles of operation of a high performance liquid chromatography system.
- To be aware of the range of detectors that can be used for high performance liquid chromatography.

Bioavailability, Bioaccessibility and Mobility of Environmental Contaminants   John R. Dean
© 2007 John Wiley & Sons, Ltd

- To appreciate the requirements for coupling a chromatography (GC and HPLC) system to a mass spectrometer.
- To appreciate the range of mass spectrometers used.

3.1 Introduction

The quantitative analysis of persistent organic pollutants (POPs) is normally carried out by using chromatographic techniques (e.g. gas chromatography or high performance liquid chromatography). However, first it is necessary to consider the different sample preparation approaches that may be required prior to analysis.

3.2 Sample Preparation for Persistent Organic Pollutant Analysis

3.2.1 Solid Samples

A range of approaches are available for the extraction of POPs from solid or semi-solid matrices.

SAQ 3.1

The extraction techniques for the recovery of POPs from (semi)-solid samples can be categorized in a variety of ways. Can you suggest a way to categorize them?

3.2.1.1 Soxhlet Extraction

A Soxhlet extraction apparatus consists of a solvent reservoir, an extraction body, an isomantle for heating and a water-cooled reflux condenser (Figure 3.1). The sample (e.g. 10 g soil) and anhydrous sodium sulfate (10 g) are placed in the porous cellulose thimble which is located in the inner tube of the Soxhlet apparatus. The latter is then fitted to a round-bottomed flask, containing the organic solvent, and to a reflux condenser. The solvent is then boiled gently using the isomantle.

DQ 3.1

What happens when the organic solvent is boiled?

Answer

When the organic solvent is boiled, the solvent vapour passes up through the tube marked (A) (Figure 3.1(a)), is condensed by the reflux condenser, and the condensed solvent falls into the thimble and slowly fills

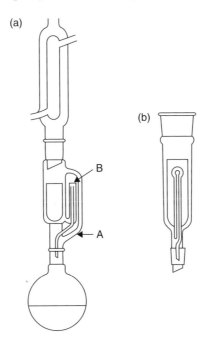

Figure 3.1 Schematic representations of two variations of a Soxhlet apparatus: (a) the solvent vapour is allowed to cool by passing to the outside of the apparatus – this extraction process is relatively slow; (b) the solvent vapour remains within the body of the apparatus – this allows more rapid extraction. From Dean, J. R., *Extraction Methods for Environmental Analysis*. Copyright 1998. © John Wiley & Sons, Limited. Reproduced with permission.

the body of the Soxhlet apparatus. When the solvent reaches the top of the tube (B), it syphons over the organic solvent containing the analyte extracted from the sample in the thimble into the round-bottomed flask. The whole process is repeated for time-periods between 6 and 24 h.

A procedure for Soxhlet extraction is shown in Figure 3.2.

3.2.1.2 Shake-Flask Extraction

Liquid–solid extraction (shake-flask extraction) is carried out by placing a sample into a suitable glass container, adding an organic solvent, and then agitating or shaking for a pre-specified time-period. After extraction, the solvent containing the POP needs to be separated from the sample matrix.

DQ 3.2

How can the solvent be separated from the sample matrix?

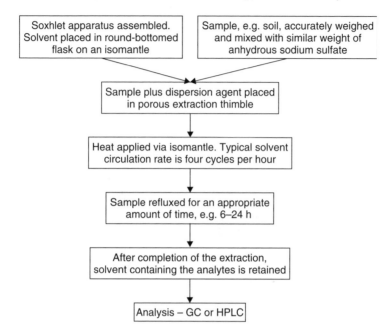

Figure 3.2 Typical procedure used for the Soxhlet extraction of solids. From Dean, J. R., *Methods for Environmental Trace Analysis*, AnTS Series. Copyright 2003. © John Wiley & Sons, Limited. Reproduced with permission.

Answer

This can be done by means of centrifugation and/or filtration.

It is advisable to repeat the process several times with fresh solvent and then combine all extracts. A procedure for shake-flask extraction is shown in Figure 3.3.

3.2.1.3 Ultrasonic Extraction

Sonication involves the use of sound waves to agitate the sample which is immersed in an organic solvent for 3 min.

DQ 3.3

How are sound waves generated in a laboratory?

Answer

This is achieved via a sonic probe or bath.

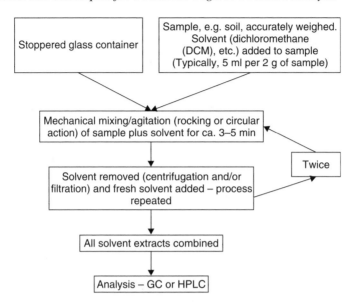

Figure 3.3 Typical procedure used for the shake-flask extraction of solids. From Dean, J. R., *Methods for Environmental Trace Analysis*, AnTS Series. Copyright 2003. © John Wiley & Sons, Limited. Reproduced with permission.

After extraction, the solvent containing the POP is separated by centrifugation and/or filtration and fresh solvent added. The whole process is repeated (three times in total) and all of the solvent extracts combined. A procedure for ultrasonic extraction is shown in Figure 3.4.

3.2.1.4 Supercritical Fluid Extraction

Supercritical fluid extraction (SFE) relies on the diversity of properties exhibited by the supercritical fluid (typically carbon dioxide) to (selectively) extract analytes from the matrix (see also Chapter 5, Section 5.2.3). A typical system consists of six basic components, i.e. a cylinder of CO_2, organic modifier, two pumps, an oven for the extraction cell, pressure outlet or restrictor and a collection vessel (see Figure 5.15 in Chapter 5).

DQ 3.4

Why are two pumps required?

Answer

Two pumps are required to pump the pressurized CO_2 and organic modifier (composition varies between 1 and 20 vol%) through the SFE system.

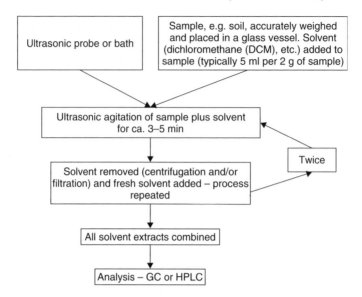

Figure 3.4 Typical procedure used for the ultrasonic extraction of solids. From Dean, J. R., *Methods for Environmental Trace Analysis*, AnTS Series. Copyright 2003. © John Wiley & Sons, Limited. Reproduced with permission.

The sample cell is located in an oven (capable of operating at temperatures of up to 200–250°C). A variable (mechanical or electronically controlled) restrictor is used to maintain the pressure within the extraction cell. A typical procedure for supercritical fluid extraction is shown in Figure 3.5.

3.2.1.5 Microwave-Assisted Extraction

Microwave-assisted extraction (MAE) utilizes organic solvent and heat to extract organic pollutants from solid matrices.

SAQ 3.2

How does microwave heating work?

Commercial microwave-assisted extraction systems allow multiple vessels to be heated simultaneously. The microwave energy output of a typical system may be 1500 W at a frequency of 2450 MHz at 100% power. Pressures (up to 800 psi) at temperatures (up to 300°C) are achievable. Typical organic solvent systems used include hexane–acetone and dichloromethane.

DQ 3.5

Why is hexane not just used on its own?

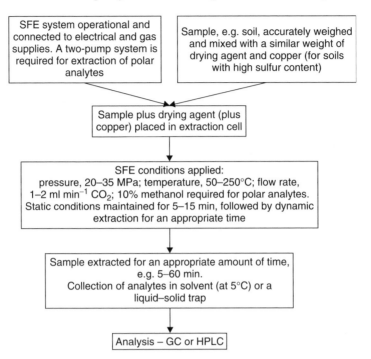

Figure 3.5 Typical procedure used for the supercritical fluid extraction of solids. From Dean, J. R., *Methods for Environmental Trace Analysis*, AnTS Series. Copyright 2003. © John Wiley & Sons, Limited. Reproduced with permission.

Answer

As hexane has no dipole moment, it will not be heated by the action of microwaves.

A typical operating procedure for the extraction of a sample of soil is shown in Figure 3.6.

3.2.1.6 Pressurized Fluid Extraction

Pressurized fluid extraction (PFE) uses heat and pressure to extract analytes rapidly and efficiently from solid matrices.

SAQ 3.3

What advantages might heat and pressure offer to potentially enhance the recovery of POPs from solid matrices?

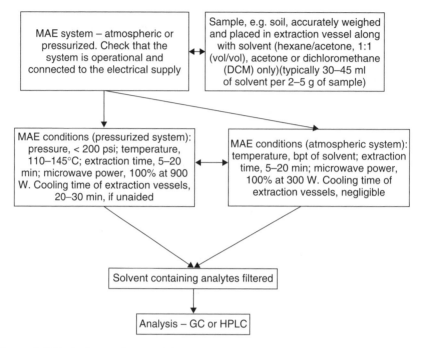

Figure 3.6 Typical procedure used for the microwave-assisted extraction of solids. From Dean, J. R., *Methods for Environmental Trace Analysis*, AnTS Series. Copyright 2003. © John Wiley & Sons, Limited. Reproduced with permission.

Pressurized fluid extraction (PFE), also known as accelerated solvent extraction (ASE™), is a fully automated sequential extraction technique.

DQ 3.6

What is the advantage of having a fully automated system?

Answer

It allows unattended operation of the PFE system once samples are pre-loaded.

The apparatus consists of a solvent-supply system, extraction cell, oven, collection system and purge system. Samples are placed in extraction cells (1, 5, 11, 22 and 33 ml). Each sample is then subjected to a temperature (40–200°C) and pressure (6.9–20.7 MPa (1000–3000 psi)) in the presence of organic solvent (acetone:dichloromethane, 1:1, vol/vol) for a short time (10 min). Then, N_2 gas (45 s at 150 psi) is used to purge both the sample cell and the stainless-steel

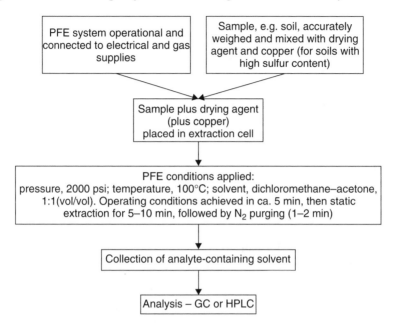

Figure 3.7 Typical procedure used for the pressurized fluid extraction (or accelerated fluid extraction) of solids. From Dean, J. R., *Methods for Environmental Trace Analysis*, AnTS Series. Copyright 2003. © John Wiley & Sons, Limited. Reproduced with permission.

transfer lines. All extracted POPs and solvent(s) are then collected in a vial. A typical procedure for pressurized fluid extraction is shown in Figure 3.7.

SAQ 3.4

What might you need to consider to identify the best extraction technique for recovery of POPs from solid samples?

3.2.2 Liquid Samples

3.2.2.1 Liquid–Liquid Extraction

The principal of liquid–liquid extraction (LLE) is that the sample is partitioned between two immiscible solvents (an organic solvent and an aqueous sample solution) in which the POP and matrix have different solubilities.

SAQ 3.5

What are the principles of liquid–liquid extraction?

3.2.2.2 Solvent Extraction

An aqueous sample (1 l, at a specified pH) is introduced into a large separating funnel (2 l capacity with a Teflon stopcock) and then a suitable organic solvent, e.g. dichloromethane (60 ml), is added. The separating funnel is then stoppered and shaken vigorously for 1–2 min. The shaking process allows thorough interspersion between the two immiscible solvents, hence maximizing the contact between the two solvent phases and assisting mass transfer, thus allowing efficient partitioning to occur.

DQ 3.7

How can the excess pressure generated be alleviated?

Answer

It is necessary to periodically vent the excess pressure generated during the shaking process by removing the stopper.

After a suitable 'resting period' (e.g. 10 min), the organic solvent is retained in a collection vessel. Fresh organic solvent is then added to the separating funnel and the process repeated again. This should be done at least three times in total. All organic extracts should be combined prior to analysis. A typical procedure for LLE, using a separating funnel, is described in Figure 3.8.

3.2.2.3 Solid-Phase Extraction

Solid-phase extraction (SPE) involves bringing an aqueous sample into contact with a solid phase or sorbent (available in discs or cartridges) whereby the POP is preferentially adsorbed onto the surface of the solid phase.

SAQ 3.6

What is the purpose of SPE?

The method of operation can be divided into five steps.

DQ 3.8

What are the five steps of SPE operation?

Answer

These are as follows: (1) wetting the sorbent, (2) conditioning of the sorbent, (3) loading of the sample, (4) rinsing or washing the sorbent to elute extraneous material and finally, (5) elution of the analyte of interest.

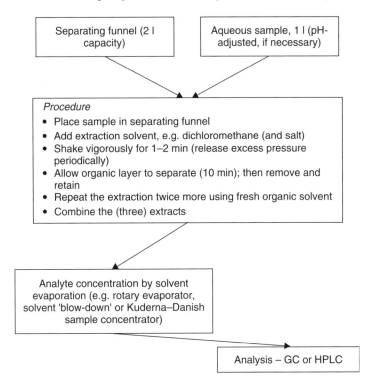

Figure 3.8 Typical procedure employed for liquid–liquid extraction, using a separating funnel. From Dean, J. R., *Methods for Environmental Trace Analysis*, AnTS Series. Copyright 2003. © John Wiley & Sons, Limited. Reproduced with permission.

A range of sorbents are available, e.g. C18 and silica, depending upon the application, and in a range of formats, e.g. cartridges or discs. A typical procedure for SPE is described in Figure 3.9.

3.2.2.4 Solid-Phase Microextraction

Solid-phase microextraction (SPME) is the process whereby an analyte is adsorbed onto the surface of a coated-silica fibre (see Figure 5.18 in Chapter 5) as a method of concentration (see also Chapter 5, Section 5.2.4). This is followed by desorption of the analytes into a suitable instrument for separation and quantitation. The most common approach for SPME is its use for gas chromatography (GC), although its coupling to high performance liquid chromatography (HPLC) has been reported. The SPME device consists of a fused-silica fibre coated with, for example, polydimethylsiloxane. SPME is incorporated into a syringe-type device, hence allowing its use in a normal unmodified injector of a gas chromatograph.

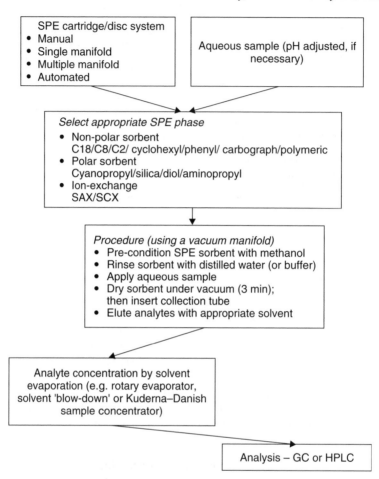

Figure 3.9 Typical procedure used for the solid-phase extraction of liquids. From Dean, J. R., *Methods for Environmental Trace Analysis*, AnTS Series. Copyright 2003. © John Wiley & Sons, Limited. Reproduced with permission.

DQ 3.9

How can SPME be used?

Answer

In operation, the SPME fibre is placed into the aqueous sample or headspace above the (solid) sample. After exposure to the POPs, the fibre is withdrawn back into its protective syringe barrel and inserted into the hot injector of the gas chromatograph. In the injector of the chromatograph, the POPs are desorbed from the fibre prior to separation and detection.

A typical procedure for SPME is described in Figure 3.10.

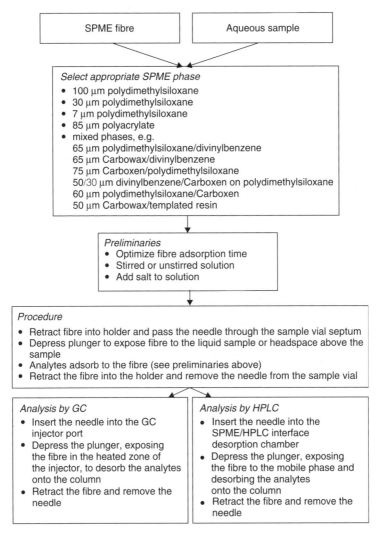

Figure 3.10 Typical procedure used for the solid-phase microextraction of liquids. From Dean, J. R., *Methods for Environmental Trace Analysis*, AnTS Series. Copyright 2003. © John Wiley & Sons, Limited. Reproduced with permission.

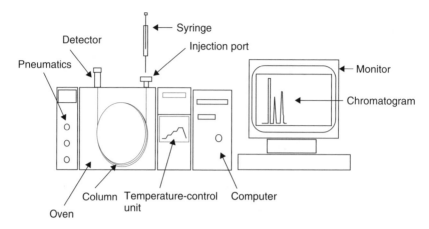

Figure 3.11 Schematic diagram of a typical gas chromatograph. Reproduced by permission of Mr E. Ludkin, Northumbria University, Newcastle, UK.

SAQ 3.7

What might you need to consider to identify the best extraction technique for recovery of POPs from liquid samples?

3.3 Gas Chromatography

Gas chromatography (GC) (Figure 3.11) can be performed under isothermal (i.e. the 'same temperature') or temperature-programmed (i.e. the temperature of the oven changing from a low temperature to a higher temperature during the chromatographic run) conditions. In capillary GC, the column (30 m long × 0.25 mm internal diameter × 0.25 μm film thickness) is located inside an oven. Samples, typically 1 μl, are introduced by using a syringe via an injector.

DQ 3.10

What is the most common injector for GC?

Answer

The most common injector is the split/splitless type (Figure 3.12) which allows a controlled volume of sample to be introduced into the column.

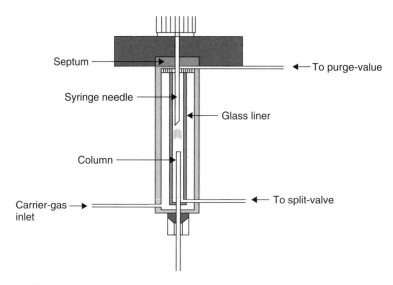

Figure 3.12 Schematic diagram of a split/splitless injector used in gas chromatography. Reproduced by permission of Mr E. Ludkin, Northumbria University, Newcastle, UK.

DQ 3.11

Why is the injector independently heated?

Answer

The injector is independently heated to allow vaporization of the POPs in an organic solvent.

A range of column types are available, including perhaps the most commonly designated 'DB-5'. This describes the chemical entity that constitutes the film which coats the internal surface of the column. A DB-5 is a low-polarity column composed of 5% diphenyl- and 95% dimethylsiloxane which is chemically bonded onto a silica support (Figure 3.13). The carrier gas, e.g. nitrogen, transports the vaporized sample through the column to the detector. The carrier gas is supplied from either a cylinder, or more commonly, a generator. A wide range of detectors are available, including universal detectors (e.g. the flame-ionization detector, Figure 3.14) through to 'more specific' detectors.

DQ 3.12

What specific detectors are available for GC?

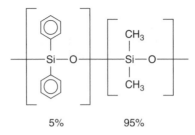

5%   95%

Figure 3.13 The stationary phase of a DB-5 gas chromatography column, consisting of 5% diphenyl- and 95% dimethylpolysiloxanes. From Dean, J. R., *Methods for Environmental Trace Analysis*, AnTS Series. Copyright 2003. © John Wiley & Sons, Limited. Reproduced with permission.

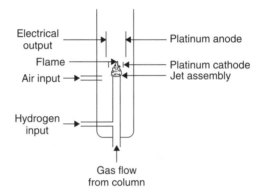

Figure 3.14 Schematic diagram of a flame-ionization detector.

Answer

These include the electron-capture detector (Figure 3.15) for detecting compounds that contain a halogen (e.g. chlorine in an organochlorine pesticide) or the nitrogen–phosphorus detector (Figure 3.16) for sensitive detection of compounds containing nitrogen and phosphorus.

The ability to couple a gas chromatograph to a mass spectrometer provides a powerful technique for identifying and quantifying POPs in the environment.

3.4 High Performance Liquid Chromatography

In contrast to GC, high performance liquid chromatography (HPLC) (Figure 3.17) is also available in two forms which characterize the technique, i.e. normal and

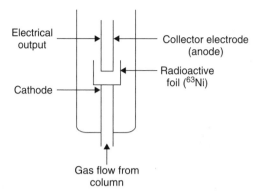

Figure 3.15 Schematic diagram of an electron-capture detector.

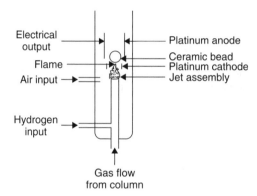

Figure 3.16 Schematic diagram of a nitrogen–phosphorus detector.

reversed-phase. In reversed-phase HPLC (RPHPLC), the column is non-polar, whereas the mobile phase is polar, e.g. a methanol–water mixture.

DQ 3.13

Having considered reversed-phase HPLC, what do you think the properties of the column and mobile phase are for normal-phase HPLC?

Answer

In normal-phase HPLC, the column is polar while the mobile phase is non-polar, e.g. heptane–isopropanol.

An HPLC system can be operated in either the *isocratic* mode, i.e. the mobile-phase composition remains constant throughout the chromatographic run, or in the *gradient* mode.

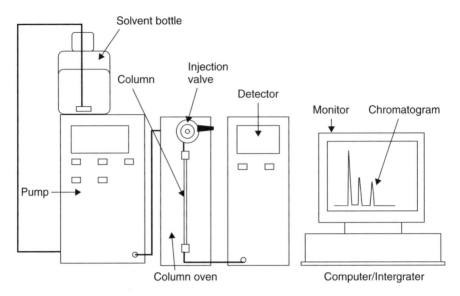

Figure 3.17 Schematic diagram of a typical isocratic high performance liquid chromatograph. Reproduced by permission of Mr E. Ludkin, Northumbria University, Newcastle, UK.

DQ 3.14

What happens to the mobile phase composition during a gradient run?

Answer

During a gradient run, the mobile phase composition varies.

The majority of applications utilize RPHPLC. The column, typically 15–25 cm long with an internal diameter ranging from 3 to 4.6 mm, is packed with a stationary phase. A typical stationary phase for RPHPLC is octadecylsilyl, often referred to as ODS.

DQ 3.15

How is ODS commonly referred to?

Answer

ODS is commonly referred to as consisting of a C18 hydrocarbon chain bonded to silica particles of 5–10 μm diameter (Figure 3.18).

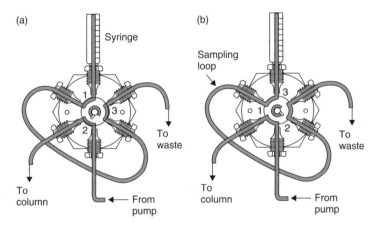

Figure 3.18 Silica particles coated with octadecylsilane (ODS) for reversed-phase high performance liquid chromatography.

Figure 3.19 Schematic diagrams of a typical injection value used for high performance liquid chromatography: (a) load position; (b) inject position. From Dean, J. R., *Methods for Environmental Trace Analysis*, AnTS Series. Copyright 2003. © John Wiley & Sons, Limited. Reproduced with permission.

The mobile phase is pumped $(1 \, ml \, min^{-1})$ through the column via a reciprocating piston pump. *En route*, the sample is introduced via an injection valve (Figure 3.19) which is capable of reproducibly introducing $5-50 \, \mu l$ of sample into the mobile phase. 'Post-separation', the eluting compounds are detected. The most common detector for HPLC is the UV/visible cell (Figure 3.20).

DQ 3.16

In which formats is a UV/visible cell available for detection in HPLC?

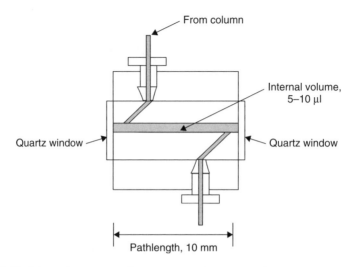

Figure 3.20 Schematic diagram of a UV/visible detector cell for high performance liquid chromatography.

Answer

This is available either as a fixed wavelength, variable wavelength or a photodiode-array detector.

DQ 3.17

What other detectors are available for detection in HPLC?

Answer

A range of other detectors are available, including fluorescence, electrochemical, refractive index, light-scattering or chemiluminescence devices.

3.5 Interfacing Chromatography and Mass Spectrometry

A major advance in chromatographic detection has been the availability of lower-cost 'bench-top' mass spectrometers.

The coupling of a capillary gas chromatograph to a mass spectrometer (Figure 3.21) is achieved via a heated transfer line which allows the separated compounds to remain in the vapour phase as they are transported in the carrier

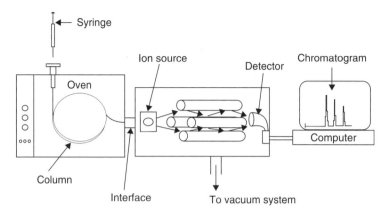

Figure 3.21 Schematic diagram of a capillary gas chromatography–mass spectrometry hyphenated system.

gas to the ionization source of the spectrometer. The preferred carrier gas for GC–MS is helium.

DQ 3.18

What are the two approaches available for ionization of compounds in GC–MS?

Answer

The approaches available for the ionization of compounds in GC–MS are chemical ionization (CI) and electron impact (EI).

Electron impact is the most common method of ionization in GC–MS. Essentially, electrons produced from a heated tungsten or rhenium filament (cathode) are accelerated towards an anode, colliding with the vaporized sample (X) and producing (positively) charged ions (Figure 3.22). These positively charged ions are then separated by the mass spectrometer. This can be expressed in the form of the following equation:

$$X_{(g)} + e^- \longrightarrow X_{(g)}{}^+ + 2e^- \tag{3.1}$$

In the CI mode, a reagent gas, e.g. methane, is required. The reagent gas is ionized by electron bombardment to produce a molecular ion ($CH_4{}^+$). This molecular ion then reacts with further methane molecules to produce a reactant ion ($CH_5{}^+$). The reactant ion then interacts with the sample molecules to produce (positively) charged ions. These ions are then separated by the spectrometer.

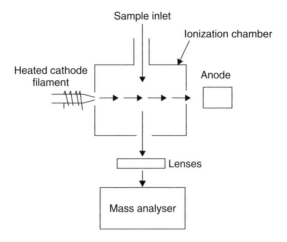

Figure 3.22 Schematic diagram of an electron-impact ionization source.

DQ 3.19

What is the major difference when using this approach?

Answer

The major difference is that the resultant (positively) charged ion has the molecular weight of the compound plus hydrogen (i.e. XH^+).

This method of ionization can be expressed as follows:

$$CH_4 + e^- \longrightarrow CH_4^+ + 2e^- \tag{3.2}$$

$$CH_4^+ + CH_4 \longrightarrow CH_5^+ + CH_3^* \tag{3.3}$$

$$X_{(g)} + CH_5^+ \longrightarrow XH_{(g)}^+ + CH_4 \tag{3.4}$$

Major developments have recently taken place such that the coupling of a liquid chromatograph to a mass spectrometer is now routine. Modern interfaces are based on the ionization of compounds at atmospheric pressure (i.e. external to the spectrometer). The two major types are electrospray (ES) ionization and atmospheric-pressure chemical ionization (APCI).

In ES ionization (Figure 3.23), the mobile phase from the liquid chromatograph is pumped through a stainless-steel capillary tube, held at a potential of between 3 and 5 kV. This results in the mobile phase being 'sprayed' from the exit of the capillary tube.

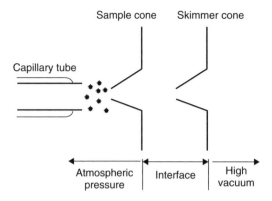

Figure 3.23 Schematic diagram of an electrospray ionization source.

DQ 3.20

What occurs as a result of the 'spraying' action?

Answer

This 'spraying' action produces highly charged solvent and solute ions in the form of droplets. Application of a continuous flow of nitrogen carrier gas allows the solvent to evaporate, so leading to the formation of solute ions.

The resultant ions are transported into the high-vacuum system of the mass spectrometer via a sample-skimmer arrangement (often positioned at a right-angle to the high-vacuum system). The formation of a potential gradient between the ES interface and the nozzle allows the generated ions to be 'pulled' into the mass spectrometer while still allowing some discrimination between the solute ions and unwanted salts (e.g. from the buffer in the mobile phase).

In atmospheric-pressure chemical ionization (APCI), the voltage is applied to a corona pin located in front of (but not in contact with) the stainless-steel capillary tubing through which the mobile phase from the chromatograph passes (Figure 3.24). To aid the process, the stainless-steel capillary tube is heated and surrounded by a coaxial flow of nitrogen gas. The interaction of the nitrogen gas and the mobile phase results in the formation of an aerosol.

DQ 3.21

How can the aerosol become desolvated?

Answer

This occurs due to the presence of heated nitrogen gas.

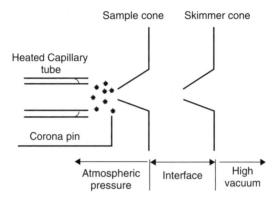

Figure 3.24 Schematic diagram of an atmospheric-pressure chemical ionization source.

A voltage (2.5–3 kV) is applied to the corona pin, hence resulting in the formation of a plasma.

DQ 3.22

What is a plasma in this situation?

Answer

A plasma is an ionized gas consisting of solute ions, solvent ions, charged carrier gas ions and electrons.

The generated ions are transported into the high-vacuum system of the mass spectrometer, as described above for ES ionization.

In these forms of ionization as used for HPLC, the molecules form singly charged ions by the loss or gain of a proton (hydrogen atom), e.g. $[M + 1]^+$ or $[M - 1]^-$, where M is the molecular weight of the compound.

DQ 3.23

What effect does this have on the operation of the mass spectrometer?

Answer

This allows the spectrometer to operate in either the positive-ion mode or negative-ion mode.

In the positive-ion mode, peaks at mass/charge ratios of M + 1, i.e. for basic compounds, e.g. amines, can be detected, whereas in the negative-ion mode peaks at mass/charge ratios of M − 1, i.e. for acidic compounds, e.g. carboxylic acids, can be detected. This process can be complicated in the presence of buffers, e.g. as part of the mobile phase, when adducts can form.

DQ 3.24

Which adducts can form as a result of adding an ammonium buffer?

Answer

This results in the occurrence of ammonium adducts with mass/charge ratios of M + 18.

DQ 3.25

Which adducts can form as a result of adding a sodium buffer?

Answer

This results in the occurrence of sodium adducts with mass/charge ratios of M + 23.

These effects can also result due to the presence of contamination in the preparation of sample/standard solutions, e.g. from the containers used to hold the solutions.

A variety of types of mass spectrometry systems are used for GC and HPLC detection. Perhaps the most common one (certainly for GC) is the quadrupole mass spectrometer, although increasingly other types of MS systems are available, including ion-trap MS and time-of-flight MS.

A quadrupole mass spectrometer (see Figure 2.24 in Chapter 2) consists of four parallel rods in which DC and RF voltages are applied to opposite rods. By changing the applied voltages, ions of a selected mass/charge ratio can pass through the spectrometer to the detector, while all other ions are lost. By continually changing the applied voltages, compounds with a range of different mass/charge ratios can pass through the spectrometer and be detected. In an ion-trap MS system (see Figure 2.25 in Chapter 2), altering the applied voltages applied to a cylindrically symmetrical electrode allows ions of increasing mass/charge ratios to be selected for detection. A time-of-flight mass spectrometer (see Figure 2.26 in Chapter 2) separates ions with respect to their velocities. This is because a charged compound accelerated, by application of a voltage, has a resultant velocity that is characteristic of its mass/charge ratio. This separation can be improved by allowing the ions to travel further. This is achieved by using a 'reflectron'. Often, a time-of-flight mass spectrometer is preceded by a quadrupole mass spectrometer. The detection of positive ions in MS requires the use of a 'horn-shaped' electron multiplier tube (EMT) (see Figure 2.27 in Chapter 2). A high potential

($-3\,kV$) is applied to the open aperture of the device. Positive ions are attracted to the high negative potential of the EMT. The semiconductor-coated internal surface of the device converts each incoming positive ion into an electron. These electrons are attracted within the 'horn-shaped' EMT to earth potential (i.e. an area that is less negative than the open aperture). As the electrons proceed within the device, they collide with the internal surface producing secondary electrons, i.e. a 'multiplication' of electrons takes place. All of the generated electrons, produced from the initial positive ion colliding with the internal surface of the device, are collected and measured. In order that both positive and negative ions can be detected in MS, the use of an electron multiplier tube with a conversion dynode prior to the normal discrete dynode is required. The conversion dynode can be segmented, i.e. one segment is coated with a material that is responsive to the negative ions while a different segment is coated with a material which is responsive to the positive ions. This allows one detector to be able to monitor in either the positive-ion or negative-ion modes.

3.6 Comparison of Persistent Organic Pollutant Analytical Techniques

Comparisons of the various analytical techniques by performing SWOT (strengths, weaknesses, opportunities and threats) analyses are shown in Tables 3.1–3.3.

Table 3.1 SWOT analysis: separation techniques (in general terms)

Strengths	• A range of organic compounds can be separated
	• Diverse range of detectors available which allow both quantitative and qualitative identification of compounds
	• Linear dynamic range of detectors is generally 'high to moderate'
Weaknesses	• Samples need to be in an appropriate form for chromatographic separation
	• Sample size/level of contamination can influence separation technique
	• Sample pre-concentration/'clean-up' often required
Opportunities	• Faster separation with shorter, more efficient columns
	• Developments in miniaturization are leading to new opportunities for separation science
Threats	• Long-term threat from electrically driven separations
	• Development of spectroscopy combined with chemometric data treatment for analysis of simple mixtures
	• Less importance of chromatographic separation with developments in mass spectrometry (MS^n)

Table 3.2 SWOT analysis: gas chromatography (GC)

Strengths	• Ability to separate a range of volatile organic compounds, i.e. polynuclear aromatic hydrocarbons (PAHs), pesticides, etc. • Wide variety of detectors available, e.g. FID, ECD, MS, etc. • Large-volume injectors allow increased sensitivity
Weaknesses	• Solid and liquid samples need to be extracted prior to analysis; pre-concentration and/or 'clean-up' often required • Optimum separation requires the most appropriate column – wide choice of columns and manufacturers available
Opportunities	• Relative low cost of GC–MS • Non-volatile compounds can be derivatized and separated by GC • Fast separations using high-temperature GC
Threats	• Potential threat from miniaturization

Table 3.3 SWOT analysis: high performance liquid chromatography (HPLC)

Strengths	• Ability to separate a range of non-volatile organic compounds, i.e. polynuclear aromatic hydrocarbons (PAHs), phenols, etc. • Wide variety of detectors available, e.g. UV/visible, fluorescence, MS, etc.
Weaknesses	• High cost of LC–MS • Solid and liquid samples need to be extracted prior to analysis; pre-concentration and/or 'clean-up' often required • Optimum separation requires the most appropriate column – wide choice of columns and manufacturers available
Opportunities	• Coupling to MS allows complex samples to be analysed by using MS^n • Fast separations using short, narrow-bore columns
Threats	• Potential threat from capillary electrophoresis and miniaturization

3.7 Selected Resources on Persistent Organic Pollutant Analytical Techniques

3.7.1 Specific Books on Chromatography[†]

Grob, K., *Split and Splitless Injection in Capillary Gas Chromatography*, 4th Edition, John Wiley & Sons, Ltd., Chichester, UK, 2001.

[†] Arranged in chronological order.

Kleibohmer, W. (Ed.), *Handbook of Analytical Separations*, Vol. 3, *Environmental Analysis*, Elsevier, Amsterdam, The Netherlands, 2001.

Ahuja, S., *Handbook of Bioseparations*, Academic Press, London, UK, 2000.

Fritz, J. S. and Gjerde, D. T., *Ion Chromatography*, 3rd Edition, John Wiley & Sons, Ltd, Chichester, UK, 2000.

Hahn-Deinstrop, E., *Applied Thin Layer Chromatography*, John Wiley & Sons Ltd, Chichester, UK, 2000.

Hubschmann, H. J., *Handbook of GC/MS*, John Wiley & Sons, Ltd, Chichester, UK, 2000.

Kromidas, S., *Practical Problem Solving in HPLC*, John Wiley & Sons, Ltd, Chichester, UK, 2000.

Subramanian, G., *Chiral Separation Techniques*, 2nd Edition, John Wiley & Sons, Ltd, Chichester, UK, 2000.

Weinberger, R., *Practical Capillary Electrophoresis*, 2nd Edition, Academic Press, London, UK, 2000.

Wilson, I., *Encyclopedia of Separation Science*, Academic Press, London, UK, 2000.

Hanai, T., *HPLC: A Practical Guide*, The Royal Society of Chemistry, Cambridge, UK, 1999.

Jennings, W., Mittlefehld, E. and Stremple, P., *Analytical Gas Chromatography*, 2nd Edition, Academic Press, London, UK, 1997.

Kolb, B. and Ettre, L. S., *Static Headspace–Gas Chromatography. Theory and Practice*, Wiley-Interscience, New York, NY, USA, 1997.

Synder, L. R., Kirkland, J. J. and Glajch, J. L., *Practical HPLC Method Development*, 2nd Edition, John Wiley & Sons Ltd, New York, NY, USA, 1997.

Weston, A. and Brown, P., *HPLC and CE: Principles and Practice*, Academic Press, London, UK, 1997.

Fowlis, I. A., *Gas Chromatography*, 2nd Edition, ACOL Series, John Wiley & Sons, Ltd, Chichester, UK, 1995.

Robards, K., Haddard, P. R. and Jackson, P. E., *Principles and Practice of Modern Chromatographic Methods*, Academic Press, London, UK, 1994.

Braithwaite, A. and Smith, F. J., *Chromatographic Methods*, 4th Edition, Chapman & Hall, London, UK, 1990.

Schoenmakers, P. J., *Optimization of Chromatographic Selectivity. A Guide to Method Development*, Elsevier, Amsterdam, The Netherlands, 1986.

Synder, L. R. and Kirkland, J. J., *Introduction to Modern Liquid Chromatography*, 2nd Edition, John Wiley & Sons, Inc., New York, NY, USA, 1979.

Summary

The different sample preparation techniques applied to solid and liquid samples for persistent organic pollutant analysis are described. In addition, the various analytical techniques capable of performing quantitative analysis of persistent organic pollutants in environmental samples are reviewed. Finally, SWOT (strengths, weaknesses, opportunities and threats) analyses are described in order to highlight the parameters that should be considered when choosing between the different analytical techniques.

Chapter 4

Methods Used to Assess Bioavailability of Metals

Learning Objectives

- To appreciate the role that non-exhaustive extraction techniques have in the study of metal bioavailability.
- To be aware of the use and procedures associated with a range of single extraction methods for metal bioavailability studies, including the use of acetic acid, ammonium nitrate, calcium chloride, ethylenediamine tetraacetic acid (EDTA), diethylenetriamine pentaacetic acid (DTPA) and sodium nitrate.
- To be aware of the use and procedure for the sequential extraction of metals in bioavailability studies.
- To appreciate the role that bioassays have in the study of metal bioavailability.
- To be aware of the use and procedures associated with the use of earthworms for assessing metal bioavailability.
- To be aware of the use and procedures associated with the use of plants for assessing metal bioavailability.
- To identify the metal-containing Certified Reference Materials available for single and sequential extraction methods.

4.1 Non-Exhaustive Extraction Techniques for Metals

Metals are normally released from solid matrices, e.g. from polluted soil, using concentrated acids and heat (often termed acid digestion – see also Section 2.2

Bioavailability, Bioaccessibility and Mobility of Environmental Contaminants   John R. Dean
© 2007 John Wiley & Sons, Ltd

(Chapter 2)). In this situation, the aim is to destroy the matrix to release the metals into solution. However, to assess the risk from metals in the environment requires the use of selective chemical extractants that can liberate metals from the soil matrix by overcoming specific interactions.

DQ 4.1

Which types of interactions need to be overcome to liberate metals from a soil matrix?

Answer

One example of the interactions that need to be overcome is ion-exchange.

This information is particularly important in environmental risk management as it can allow a prediction of the potential likelihood of metal release, transformation, mobility or availability as the soils are exposed to weathering, pH changes and changes in land-use. A range of different approaches are possible, based on the use of particular reagents, that seek to isolate and/or identify the particular soil phase in which the metal is retained/trapped. These approaches are identified in Table 4.1. The variety of approaches available meant that, initially, it was difficult to identify the specific approach required for a particular task. Therefore the Standard, Measurements and Testing Programme (SM & T – formerly BCR) of the European Union [2–4] identified a series of approaches that could be used for single or sequential extraction of metals from soil matrices.

4.2 Single Extraction Methods for Metals

The main procedures identified for the extraction of metals using single extraction methods are based on the use of ethylenediamine tetraacetic acid (EDTA), acetic acid or diethylenetriamine pentaacetic acid (DTPA). The procedures based on solutions of EDTA ($0.05\,mol\,l^{-1}$), acetic acid ($0.43\,mol\,l^{-1}$) or DTPA ($0.005\,mol\,l^{-1}$ DTPA) are shown in Figures 4.1–4.3, respectively.

When performing single extraction procedures (Figures 4.1–4.3), it is important to consider the following general laboratory practices:

- All laboratory ware should be made of borosilicate glass, polypropylene, polyethylene or polytetrafluoroethylene (PTFE), except for the centrifuge tubes which should be made of borosilicate glass or PTFE.

- All vessels in contact with samples or reagents should be cleaned in HNO_3 ($4\,mol\,l^{-1}$) for at least 30 min, then rinsed with distilled water, cleaned with $0.05\,mol\,l^{-1}$ EDTA and rinsed again with distilled water. Alternatively, clean

Table 4.1 Extraction procedures used to isolate nominal soil/sediment phases [1]

Phase	Reagent or method of isolation	Comments
Water-soluble, soil solution, sediment pore water	Water, centrifugation, displacement, filtration, dialysis	Contains the most mobile and hence potentially available metal species
Exchangeable	$MgCl_2$, NH_4OAc, HOAc	Contains weakly bound (electrostatically) metal species that can be released by ion-exchange with cations such as Ca^{2+}, Mg^{2+} or NH_4^+. Ammonium acetate is the preferred extractant as the complexing power of acetate prevents re-adsorption or precipitation of released metal ions. In addition, acetic acid dissolves the exchangeable species, as well as more tightly bound exchangeable forms
Organically bound	$Na_4P_2O_7$, H_2O_2 at pH 3/NaOAC	Contains metals bound to the humic material of soils. Sodium hypochlorite is used to oxidize the soil organic matter and release the bound metals. An alternative approach is to oxidize the organic matter with 30% hydrogen peroxide, acidified to pH 3, followed by extraction with ammonium acetate to prevent metal ion re-adsorption or precipitation
Carbonate	NaOAc at pH 5 (HOAc)	Contains metals that are dissolved by sodium acetate acidified to pH 5 with acetic acid
Mn oxide-bound	$NH_2OH.HCl$	Acidified hydroxylamine hydrochloride releases metals from the manganese oxide phase with minimal attack on the iron oxide phases
Fe (amorphous) oxide	$(NH_4)_2C_2O_4$ in the dark	Amorphous forms of iron oxides can be discriminated between by extracting with acid ammonium oxalate in the dark
Fe (crystalline) oxide	$(NH_4)_2C_2O_4$ under UV light	Crystalline forms of iron oxides can be discriminated between by extracting with acid ammonium oxalate under UV light

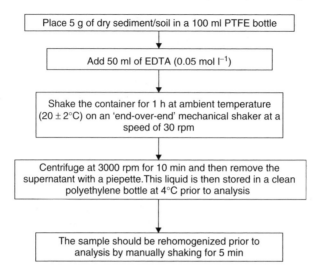

Figure 4.1 Procedure adopted in the single extraction method for metals (employing ethylenediamine tetraacetic acid (EDTA)), as applied to the analysis of soils and sediments [45]. From Dean, J. R., *Methods for Environmental Trace Analysis*, AnTS Series. Copyright 2003. © John Wiley & Sons, Limited. Reproduced with permission.

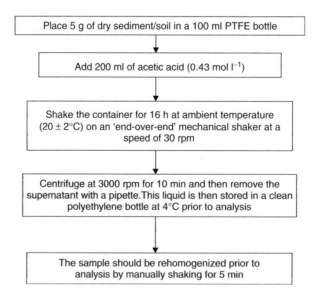

Figure 4.2 Procedure adopted in the single extraction method for metals (employing acetic acid), as applied to the analysis of soils and sediments [45]. From Dean, J. R., *Methods for Environmental Trace Analysis*, AnTS Series. Copyright 2003. © John Wiley & Sons, Limited. Reproduced with permission.

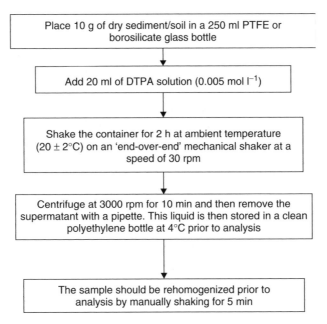

Figure 4.3 Procedure adopted in the single extraction method for metals (employing diethylenetriamine pentaacetic acid (DTPA)), as applied to the analysis of soils and sediments [45]. From Dean, J. R., *Methods for Environmental Trace Analysis*, AnTS Series. Copyright 2003. © John Wiley & Sons, Limited. Reproduced with permission.

all vessels by immersing in HNO_3 ($4 \, mol \, l^{-1}$) overnight and then rinse with distilled water.

- The mechanical shaker, preferably of the 'end-over-end' type, should be operated at a speed of 30 rpm. All samples should be centrifuged at 3000 g.

The three reagents should be prepared as follows:

- *EDTA ($0.05 \, mol \, l^{-1}$)*. In a fume cupboard, add $146 \pm 0.05 \, g$ of EDTA (free acid) to $800 \pm 20 \, ml$ of distilled water. To aid dissolution of EDTA, stir in $130 \pm 5 \, ml$ of saturated ammonia solution (prepared by bubbling ammonia gas into distilled water). Continue to add the ammonia solution until all of the EDTA has dissolved. The resultant solution should be filtered, if necessary, through a filter paper of porosity $1.4–2.0 \, \mu m$ into a pre-cleaned 10 l polyethylene bottle and then diluted to $9.0 \pm 0.5 \, l$ with distilled water. Adjust the pH to 7.00 ± 0.05 by addition of a few drops of either ammonia or hydrochloric acid, as appropriate. The solution should then be made up to 10 l with distilled water to obtain an EDTA solution of $0.05 \, mol \, l^{-1}$. Analyse a sample of each fresh batch of EDTA solution for its metal impurity content.

- *Acetic acid (0.43 mol l^{-1}).* In a fume cupboard, add 250 ± 2 ml of glacial acetic acid ('AnalaR' or similar) to approximately 5 l of distilled water in a pre-cleaned 10 l polyethylene bottle and make up to 10 l with distilled water. Analyse a sample of each fresh batch of acetic acid solution for its metal impurity content.

- *DTPA (0.005 mol l^{-1}).* In a fume cupboard, dissolve 149.2 g triethanolamine (0.01 mol l^{-1}), 19.67 g DTPA (0.005 mol l^{-1}) and 14.7 g calcium chloride in approximately 200 ml of distilled water. Allow the DTPA to dissolve and then dilute to 9 l. Adjust the pH to 7.3 ± 0.5 with HCl while stirring and then dilute to 10 l in distilled water. Analyse a sample of each fresh batch of DTPA solution for its metal impurity content.

The use of either reagent represents a suitable single extraction method for the following trace metals: Cd, Cr, Cu, Ni, Pb and Zn.

Note: When using single extraction methods for the analysis of sediment or soil samples, a separate sub-sample should be dried (in a layer of ca. 1 mm depth) in an oven at $105 \pm 2°C$ for 2–3 h, transferred to a desiccator and allowed to cool prior to weighing.

DQ 4.2

Why is it necessary to dry a sub-sample of the soil or sediment?

Answer

Drying a sub-sample allows all analytical results to be corrected to a 'dry-mass' basis, i.e. a quantity per g dry sediment/soil.

Complete experimental details for the single extraction methods are shown in Figures 4.1–4.3. However, it is important to take note of the following:

- Calibration solutions should be prepared with the appropriate extraction solution.

DQ 4.3

What is the term used for this type of calibration solution?

Answer

The term used is *matrix-matched* calibration solution.

- It is important to prepare a sample blank for every batch of extractions.

DQ 4.4

How is a sample blank prepared?

Answer

A sample blank is prepared in the same manner as the sample, except that no sediment/soil is added.

- Ensure that the sample, i.e. sediment/soil, does not form a 'cake' during the extraction procedure.

DQ 4.5

If a 'cake' is formed during the extraction procedure, what can be done?

Answer

If this occurs, either adjust the shaking speed to ensure that the suspension is maintained or mechanically break the solid 'cake' with a pre-cleaned glass rod. It is important that the sample remains in complete suspension during the extraction process.

- It is recommended for ICP–AES/MS that the extracts are filtered (0.45 μm) prior to analysis.

DQ 4.6

Why is it necessary to filter a sample for ICP analysis?

Answer

It is necessary so that the nebulizer used for sample introduction into an ICP is not blocked.

SAQ 4.1

How might you unblock a nebulizer?

For GFAAS, it is recommended that the *standard-additions method* of calibration is used.

Results from the certification of two Certified Reference Materials using EDTA and acetic acid are shown in Table 4.2.

Table 4.2 Extractable metal contents of two Certified Reference Materials (CRM 483 and CRM 484) used in the analysis of soils or sediments [45]

Certified Reference Material/extract	Certified value (mg kg^{-1})	Uncertainty (mg kg^{-1})
CRM 483 (Sewage sludge-amended soil)		
EDTA extracts		
Cd	24.3	1.3
Cr	28.6	2.6
Cu	215.0	11.0
Ni	28.7	1.7
Pb	229.0	8.0
Zn	612.0	19.0
Acetic acid extracts		
Cd	18.3	0.6
Cr	18.7	1.0
Cu	33.5	1.6
Ni	25.8	1.0
Pb	2.1	0.25
Zn	620.0	24.0
CRM 484 (terra rossa soil)		
EDTA extracts		
Cd	0.51	0.03
Cu	88.1	3.8
Ni	1.39	0.11
Pb	47.9	2.6
Zn	152.0	7.0
Acetic acid extracts		
Cd	0.48	0.04
Cu	33.9	1.4
Ni	1.69	0.15
Pb	1.17	0.16
Zn	193.0	7.0
CRM 484 (terra rossa soil)		
EDTA extracts		
Cd	2.68	0.09
Cr	0.205	0.022
Cu	57.3	2.5
Ni	4.52	0.25
Pb	59.7	1.8
Zn	383.0	12.0
Acetic acid extracts		
Cd	1.34	0.04

Table 4.2 (*continued*)

Certified Reference Material/extract	Certified value (mg kg^{-1})	Uncertainty (mg kg^{-1})
Cr	0.014	0.003
Cu	32.3	1.0
Ni	3.31	0.13
Pb	15.0	0.5
Zn	142.0	6.0

In addition to the single extraction methods described above, other procedures are also available and include the use of ammonium nitrate, calcium chloride and sodium nitrate. The procedures for the preparation of each of these single extraction solutions is now described.

NH_4NO_3 (1 mol l^{-1}). In a fume cupboard, dissolve 80.04 g of NH_4NO_3 in doubly distilled water, then make up to 1 l with the same water. Analyse a sample of each fresh batch of NH_4NO_3 solution for its metal impurity content.

$CaCl_2$ (0.01 mol l^{-1}). In a fume cupboard, dissolve 1.470 g of $CaCl_2.2H_2O$ in doubly distilled water, then make up to 1 l with the same water. Verify that the Ca concentration is 400 ± 10 mg l^{-1} by EDTA titration. Analyse a sample of each fresh batch of $CaCl_2$ solution for its metal impurity content.

$NaNO_3$ (0.1 mol l^{-1}). In a fume cupboard, dissolve 8.50 g of $NaNO_3$ in doubly distilled water, then make up to 1 l with the same water. Analyse a sample of each fresh batch of $NaNO_3$ solution for its metal impurity content.

The approach adopted for each of these single extraction methods is shown in Figures 4.4–4.6.

Note: As before, a separate sub-sample should be dried (in a layer of ca. 1 mm depth) in an oven at $105 \pm 2°$ C for 2–3 h, transferred to a desiccator and allowed to cool prior to weighing. This will allow all analytical results to be corrected to a 'dry-mass' basis, i.e. a quantity per g dry sediment/soil.

DQ 4.7

What other correction is required in the case of sodium nitrate ($NaNO_3$)?

Answer

In this approach, the results need to be corrected additionally for the difference in final volume, i.e. 2 ml of HNO_3 should be added to 48 ml of extract to give a final volume of 50 ml.

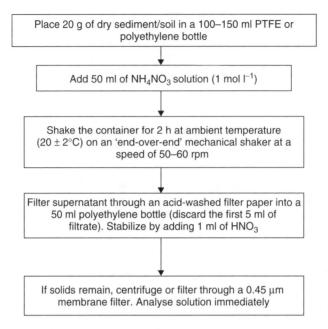

Figure 4.4 Procedure adopted in the single extraction method for metals (employing ammonium nitrate (NH_4NO_3)), as applied to the analysis of soils and sediments [2].

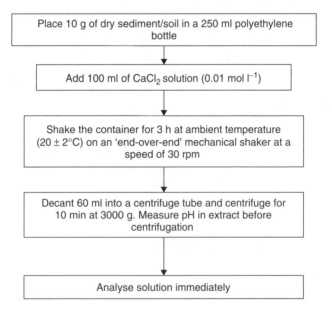

Figure 4.5 Procedure adopted in the single extraction method for metals (employing calcium chloride ($CaCl_2$)), as applied to the analysis of soils and sediments [2].

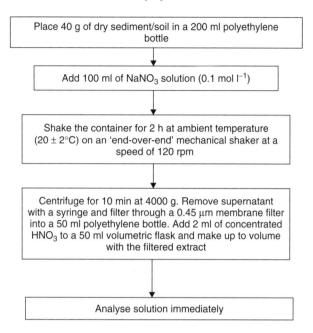

Figure 4.6 Procedure adopted in the single extraction method for metals (employing sodium nitrate (NaNO₃)), as applied to the analysis of soils and sediments [2].

Single extraction procedures using $CaCl_2$, $NaNO_3$ and NH_4NO_3 have been applied to the analysis of ten contaminated soils and a soil certified reference material [5]. The results for the extractable metal content of BCR 483 are shown in Table 4.3.

DQ 4.8

Comment on the results displayed in Table 4.3.

Answer

In all cases, except for Cu, the data obtained agreed with the indicative values reported [6].

4.3 Sequential Extraction Techniques for Metals

The procedure adopted for the sequential extraction of metals from soil/sediments is based on three distinct stages, followed by a final acid digestion to assess the

Table 4.3 Analysis of a certified reference material using various single extraction procedures[a] [5]

Material used	Cd (mg kg⁻¹)		Cu (mg kg⁻¹)		Pb (mg kg⁻¹)		Zn (mg kg⁻¹)	
	Indicative value[b]	Recovery[c]	Indicative value[b]	Recovery[c]	Indicative value[b]	Recovery[c]	Indicative value[b]	Recovery[c]
$CaCl_2$	0.45 ± 0.05	0.48 ± 0.01	1.2 ± 0.4	1.73 ± 0.06	< 0.06	< DL[d]	8.3 ± 0.7	8.1 ± 0.4
$NaNO_3$	0.08 ± 0.03	0.104 ± 0.008	0.89 ± 0.22	1.22 ± 0.07	< 0.03	< DL[d]	2.7 ± 0.8	3.3 ± 0.1
NH_4NO_3	0.26 ± 0.05	0.25 ± 0.01	1.2 ± 0.3	1.69 ± 0.07	0.020 ± 0.013	0.032 ± 0.004	6.5 ± 0.9	5.4 ± 0.3

[a]Mean concentration, ± 1 × standard deviation (SD).
[b]$n = 5$ (data from Quevauviller et al. [6]).
[c]$n = 3$.
[d]Below detection limit (DL) of ICP–MS.

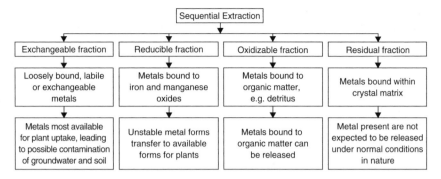

Figure 4.7 Overview of the sequential extraction method for metals, as applied to the analysis of soils and sediments [45]. From Dean, J. R., *Methods for Environmental Trace Analysis*, AnTS Series. Copyright 2003. © John Wiley & Sons, Limited. Reproduced with permission.

unavailable or residual component. An overview of the sequential extraction method is shown in Figure 4.7, and the stages are described as follows:

- *Stage 1 (Exchangeable fraction)*. The metals released in this stage are those which are most available and hence most mobile. They include those weakly absorbed metals retained on the sediment/soil surface by relatively weak electrostatic interaction, metals that can be released by ion-exchange processes and metals that can be co-precipitated with the carbonates present in many sediments/soils. Changes in the ionic composition, influencing adsorption–desorption reactions, or lowering of pH, could cause mobilization of metals from this fraction.

- *Stage 2 (Reducible fraction)*. This stage identifies those metals that are bound to iron/manganese oxides and that are therefore unstable under reduction conditions. Changes in the redox potential (E_h) could induce the dissolution of these oxides, so leading to their release from the soil/sediment.

- *Stage 3 (Oxidizable fraction)*. This stage releases those metals that are bound to organic matter within the sediment/soil matrix into solution.

- *Residual fraction*. It is common practice to 'acid-digest' by using aqua regia, the resultant sediment/soil sample and then analyse for trace metals.

SAQ 4.2

What are the constituents of aqua regia?

The residual fraction contains naturally occurring minerals which may hold trace metals within their crystalline matrix. It is the expectation of this sequential extraction procedure that these metals are not likely to be released under normal environmental conditions and hence are neither available nor likely to cause any environmental risk.

Note: The use of aqua regia to digest the residual fraction will only liberate metals that are not bound to 'silicate minerals'. As it is unlikely that the silicate-bound metals will leach from the soil/sediment, the use of aqua regia to give a 'pseudo-total' analysis is perfectively acceptable.

DQ 4.9

How can complete acid digestion of silicate-bound metals be achieved?

Answer

Complete acid digestion of silicate-bound metals can be achieved by using hydrofluoric acid.

When performing the sequential extraction method (Figures 4.8–4.10), it is important to consider the following general laboratory practices:

- All laboratory ware should be made of borosilicate glass, polypropylene, polyethylene or PTFE, except for the centrifuge tubes which should be made of borosilicate glass or PTFE.

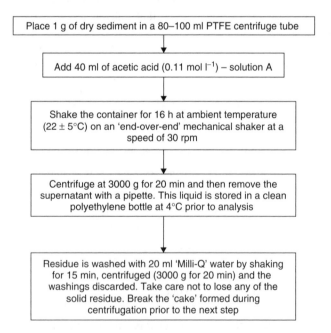

Figure 4.8 Details of Step 1 of the sequential extraction method (cf. Figure 4.7) [7].

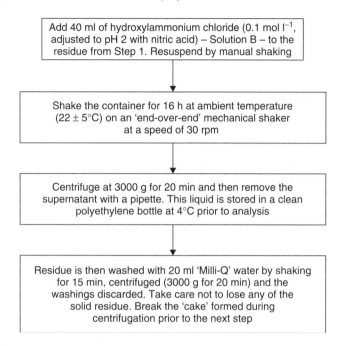

Add 40 ml of hydroxylammonium chloride (0.1 mol l^{-1}, adjusted to pH 2 with nitric acid) – Solution B – to the residue from Step 1. Resuspend by manual shaking

Shake the container for 16 h at ambient temperature (22 ± 5°C) on an 'end-over-end' mechanical shaker at a speed of 30 rpm

Centrifuge at 3000 g for 20 min and then remove the supernatant with a pipette. This liquid is stored in a clean polyethylene bottle at 4°C prior to analysis

Residue is then washed with 20 ml 'Milli-Q' water by shaking for 15 min, centrifuged (3000 g for 20 min) and the washings discarded. Take care not to lose any of the solid residue. Break the 'cake' formed during centrifugation prior to the next step

Figure 4.9 Details of Step 2 of the sequential extraction method (cf. Figure 4.7) [7].

- All vessels in contact with samples or reagents should be cleaned by immersing in HNO$_3$ (4 mol l^{-1}) overnight and then rinsing with distilled water.

- The mechanical shaker, preferably of the 'end-over-end' type, should be operated at a speed of 30 ± 10 rpm and a temperature of 22 ± 5°C. All samples should be centrifuged at 3000 g for 20 min.

The four reagents used in the three stages are prepared as follows:

- *Solution A (acetic acid, 0.11 mol l^{-1})*. Add in a fume cupboard, 25 ± 0.1 ml of glacial acetic acid ('AnalaR' or similar) to approximately 0.5 l of distilled water in a 1 l polyethylene bottle and make up to 1 l with distilled water. Take 250 ml of this solution (acetic acid, 0.43 mol l^{-1}) and dilute to 1 l with distilled water to obtain an acetic acid solution of 0.11 mol l^{-1}. Analyse a sample of each fresh batch of solution A for its metal impurity content.

- *Solution B (hydroxylamine hydrochloride or hydroxyammonium chloride, 0.5 mol l^{-1})*. Dissolve 34.75 g of hydroxylamine hydrochloride in 400 ml of distilled water. Transfer to a 1 l volumetric flask and add 25 ml of 2 mol l^{-1} HNO$_3$ (prepared by weighing from a concentration solution) (the pH should be 1.5). Make up to 1 l with distilled water. Prepare this solution on the same

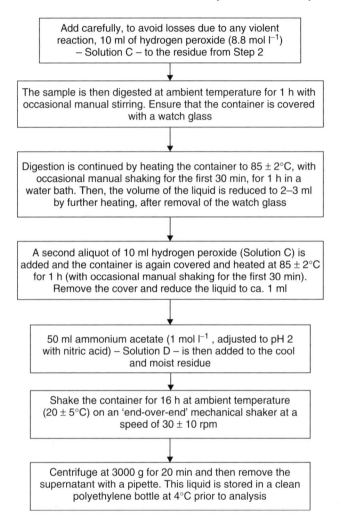

Figure 4.10 Details of Step 3 of the sequential extraction method (cf. Figure 4.7) [7].

day as the extraction is carried out. Analyse a sample of each fresh batch of solution B for its metal impurity content

- *Solution C (hydrogen peroxide, 300 mg g^{-1}, i.e. 8.8 mol l^{-1}).* Use the H$_2$O$_2$ as supplied by the manufacturer, i.e. acid-stabilized to pH 2–3. Analyse a sample of each fresh batch of solution C for its metal impurity content.

- *Solution D (ammonium acetate, 1 mol l^{-1}).* Dissolve 77.08 g of ammonium acetate in 800 ml of distilled water. Adjust to pH 2 ± 0.1 with concentrated

HNO_3 and make up to 1 l with distilled water. Analyse a sample of each fresh batch of solution D for its metal impurity content.

The use of this sequential extraction method can be applied to the following trace metals: Cd, Cr, Cu, Ni, Pb and Zn.

Note: When using this procedure for the analysis of sediment or soil samples, a separate sub-sample (1 g) should be dried (in a layer of ca. 1 mm depth) in an oven at $105 \pm 2°C$ for 2–3 h, transferred to a desiccator and allowed to cool prior to weighing. This will allow all analytical results to be corrected to a 'dry-mass' basis, i.e. a quantity per g dry sediment/soil.

Complete experimental details for the modified sequential extraction method [7] are shown in Figures 4.8–4.10. The residue from Step 3 should be transferred to a digestion vessel (see below) with approximately 3 ml of water and digested as shown in Figure 4.11. Results from the residue digest should be summed with the metal contents obtained from Steps 1–3. The result of this approach should be compared to the result obtained by aqua regia digestion of a separate, 1 g sample of the sediment/soil. The procedure for the aqua regia extraction approach to determine the elemental composition of the residual fraction and that to be followed for a fresh sample is shown in Figure 4.11.

The reagents used for aqua regia digestion are as follows:

- Hydrochloric acid – concentrated HCl = $12.0 \, mol \, l^{-1}$, $\rho = 1.19 \, g \, ml^{-1}$.

- Nitric acid – concentrated HNO_3 = $15.8 \, mol \, l^{-1}$, $\rho = 1.42 \, g \, ml^{-1}$.

- Nitric acid ($0.5 \, mol \, l^{-1}$) – dilute 32 ml of concentrated HNO_3 with water to 1 l.

In performing this sequential extraction procedure, it is important to take note of the following:

- Calibration solutions should be prepared with the appropriate extraction solutions, i.e. use 'matrix-matched' calibration solutions.

- It is important to prepare a sample blank for every batch of extractions, i.e. prepare a container with no sediment/soil, but treated in the same manner as though it contained the sample.

- Ensure that the sample, i.e. sediment/soil, does not form a 'cake' during the extraction procedure. If this occurs, either adjust the shaking speed to ensure that the suspension is maintained or mechanically break the solid 'cake' with a pre-cleaned glass rod. It is important that the sample remains in complete suspension during the extraction process.

- It is recommended for ICP–AES/MS that the extracts are filtered ($0.45 \, \mu m$) prior to analysis to prevent nebulizer blockages. For GFAAS, it is recommended that the *standard-additions method* of calibration is used.

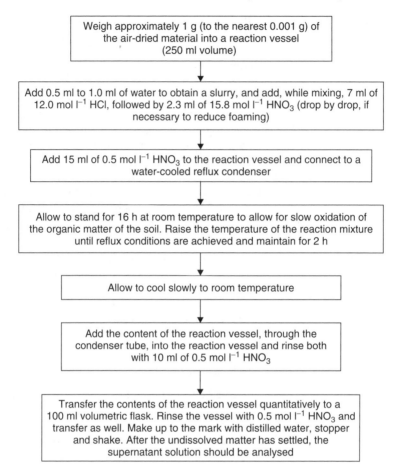

Figure 4.11 Details of the digestion process employed for the residual fraction in the sequential extraction method (cf. Figure 4.7) [7].

Results for the sequential extraction of a certified reference material by using the modified approach are shown in Table 4.4.

DQ 4.10

Comment on the results displayed in Table 4.4.

Answer

The results (Table 4.4) show the recoveries from each step of the modified (European) Community Bureau of Reference (BCR) sequential extraction (three-step) method for Cd, Cr, Cu, Ni, Pb and Zn. In addition,

Table 4.4 Comparison of the sequential extraction procedure with aqua regia-extraction protocols [7]

Stage, etc.	Element (mg kg^{-1})					
	Cd	Cr	Cu	Ni	Pb	Zn
Step 1	10.0 ± 0.77	9.40 ± 3.5	16.8 ± 1.5	17.9 ± 2.0	0.756 ± 0.70	441 ± 39
Step 2	24.8 ± 2.3	654 ± 108	141 ± 20	24.4 ± 3.3	379 ± 21	438 ± 56
Step 3	1.22 ± 0.48	2215 ± 494	132 ± 29	5.9 ± 1.4	66.5 ± 22	37.1 ± 9.9
Residual fraction	0.423 ± 0.16	183 ± 40	43.3 ± 3.8	15.2 ± 4.3	76.9 ± 17	82.1 ± 9.6
Total^a	36.44 ± 2.5	3061 ± 507	333 ± 35	63.4 ± 5.9	523 ± 35	998 ± 70
Aqua regia^b	36.40 ± 2.8	3392 ± 484	362 ± 12	63.8 ± 7.7	501 ± 47	987 ± 37

^{a} Sum of Steps 1–3 and residual fraction.
^{b} 'Pseudo-total' metal content determined by using aqua regia extraction.

data are presented for each element recovered in the residual fraction. The total recovery for each element is then summed for Steps 1–3, plus the residual fraction component, and compared to the recovery obtained when the 'sludge-amended' soil was digested with aqua regia only ('pseudo-total'). It is noted that the relative error is always < 10% and in the majority of cases is < 5%.

The modified BCR sequential extraction (three-step) method has been applied by, for example, Mossop and Davidson [8], to the analysis of a range of soils and sediments. These authors report results for samples that included a freshwater sediment, a sewage 'sludge-amended' soil, two industrial soils and an 'inter-tidal' sediment. The results for iron and manganese in the freshwater sediment, sewage 'sludge-amended' soil and an industrial soil are shown in Table 4.5. All extracts were analysed by using inductively coupled plasma–atomic emission spectroscopy. In addition, two Certified Reference Materials were analysed by using the modified BCR sequential extraction (three-step) method for Cu, Pb and Zn (Table 4.6). It was found that most results were within the indicative ranges indicated for Cu, Pb and Zn. Table 4.6 highlights in italics those data that are outside of the standard deviation expected, according to the indicative values.

4.4 Earthworms

Over 5500 named species of earthworm exist worldwide [11] and range in size from 2 cm to 3 m. One of the most common earthworm species are the reddish coloured *Lumbricus terrestris* which are commonly found in temperate regions. The main families of earthworms and their geographic origins are as follows [11]:

• Lumbricidae: temperate areas of Northern Hemisphere, mostly Eurasia

• Hormogastridae: Europe

Table 4.5 Analysis of soil and sediment samples using the modified BCR sequential extraction (three-step) procedure.[a] $n = 5$ (unless otherwise indicated) [8]

Stage, etc.	Fe (mg kg⁻¹ dry weight)			Mn (mg kg⁻¹ dry weight)		
	Freshwater sediment	Sewage 'sludge-amended' soil	Industrial soil	Freshwater sediment	Sewage 'sludge-amended' soil	Industrial soil
Step 1	209 (30.1)	32.8 (3.35)	25.6 (12.5)	28.5 (2.81)	120 (2.50)	45.8 (14.2)
Step 2	2320 (2.68)	7000 (2.63)[b]	4280 (2.71)	15.4 (3.25)	95.7 (8.88)	96.9 (9.60)
Step 3	676 (14.6)	724 (19.5)	1060 (18.2)	9.13 (8.76)	7.90 (7.34)	23.6 (17.2)
Residual fraction	12 800 (15.2)	15 340 (3.55)	60 800 (15.8)	98.3 (13.7)	41.9 (8.83)	392 (15.0)[c]
Total of 3 steps plus residual	16 005 (12.2)	23 097 (2.56)	66 166 (14.5)	151 (9.27)	266 (3.76)	558 (10.8)
'Pseudo-total'	14 300[d]	22 700[d]	46 600[d]	109[d]	220[d]	370[d]
Recovery (%)	112	102	142	139	121	151

[a]Mean – % relative standard deviations (% RSDs) shown in parentheses.
[b]$n = 4$.
[c]$n = 2$.
[d]$n = 3$.

Table 4.6 Analysis of two Certified Reference Materials using the modified BCR sequential extraction (three-step) procedure[a] [8]

	Cu (mg kg⁻¹ dry weight)		Pb (mg kg⁻¹ dry weight)		Zn (mg kg⁻¹ dry weight)	
	Indicative value	Recovery	Indicative value	Recovery	Indicative value	Recovery
CRM601[b]						
Step 1	10.5 ± 0.8^c	9.6 ± 0.4^c	2.28 ± 0.44^c	2.42 ± 1.70^c	261 ± 13^c	211 ± 9.0^c
Step 2	72.8 ± 4.9^c	48 ± 6.0^c	205 ± 11^c	182 ± 11^c	266 ± 17^c	240 ± 10^c
Step 3	78.6 ± 8.9^c	93 ± 4.0^c	19.7 ± 5.8^c	23.1 ± 2.0^c	106 ± 11^c	135 ± 8.0^c
Residual	60.4 ± 4.9^d	57 ± 8.0^d	38.0 ± 8.7^d	53.2 ± 7.5^c	161 ± 14^d	167 ± 19^c
CRM483[e]						
Step 1	16.8 ± 1.5^d	17.6 ± 0.3^e	0.76 ± 0.70^d	$<0.3^e$	441 ± 39^d	453 ± 9.0^e
Step 2	141 ± 20^d	123 ± 3.7^f	379 ± 21^d	329 ± 6.0^f	438 ± 56^d	433 ± 22^f
Step 3	132 ± 29^d	143 ± 8.0^f	66.5 ± 22^d	27.1 ± 4.6^f	37.1 ± 9.9^d	36.3 ± 5.6^f
Residual	43.3 ± 3.8^d	40.6 ± 2.3^f	76.9 ± 17^d	131 ± 5.0^f	82.1 ± 9.6^d	78.1 ± 9.6^f

[a] Mean – ± 1× standard deviation (SD).
[b] Indicative values from Rauret *et al.* [9].
[c] $n = 7$.
[d] $n = 6$.
[e] Indicative values from Rauret *et al.* [10].
[f] $n = 5$.

- Sparganophilidae: North America

- Almidae: Africa, South America

- Megascolecidae: South-East Asia, Australia and Oceania, western North America

- Acanthodrilidae: Africa, south-eastern North America, Central and South America, Australia and Oceania

- Ocnerodrilidae: Central and South America, Africa

- Octochaetidae: Central America, India, New Zealand, Australia

- Exxidae: Central America

- Glossoscolecidae: Central and northern South America

- Eudrilidae: Africa and South Africa

Earthworms have a closed circulatory system (Figures 4.12 and 4.13), consisting of two main blood vessels, i.e. the ventral blood vessel which leads the blood to

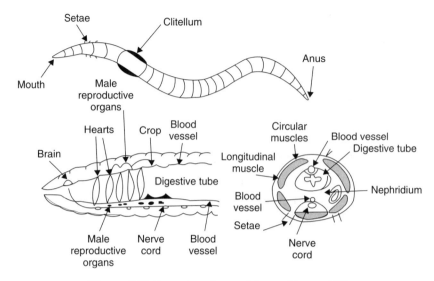

Figure 4.12 General anatomy of the earthworm [11].

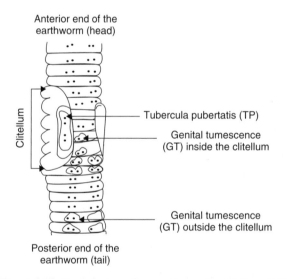

Figure 4.13 Ventral view of an earthworm's clitellum [11].

the posterior end and the dorsal blood vessel which leads to the anterior end [11]. The dorsal vessel is contractile and pumps blood forward into the ventral vessel by a series of 'hearts' (i.e. aortic arches) which vary in number in the different taxa. A typical lumbricid will have five pairs of hearts, i.e. a total of ten hearts. The blood is distributed from the ventral vessel into capillaries on the body wall and other organs and into a vascular sinus in the gut wall where gases and nutrients are exchanged. This arrangement may be complicated in the various groups by suboesophageal, supraoesophageal, parietal and neural vessels, but the basic arrangement holds in all earthworms [11].

SAQ 4.3

How do earthworms travel underground?

In compacted soils, the earthworm eats its way through the soil. This is done by the earthworm cutting a passage with its muscular pharynx and dragging the rest of the body along. The ingested soil is ground-up, digested, and the waste deposited behind the earthworm. Figures 4.14–4.16 give a field guide to the identification of common earthworms. This identification is based first on colour (dark red, red or red–violet in Figure 4.14, muddy green in Figure 4.15 and other colours in Figure 4.16) and secondly on a diagram illustrating the ventral side of the earthworm's clitellum (see also Figure 4.13) [12–14].

4.4.1 Earthworms in Bioavailability Studies

Earthworms allow a direct biological measurement of chemicals in soil and as such are ideally suited to assess bioavailability for the following reasons [15]:

- Earthworms reside in soil and are more or less in constant contact with some portion of the soil.

Aporrectodea longa
Common name: Black head worm, *Length*: 90–150 mm

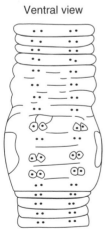

Ventral view   *Habitat*: This species has been found in cultivated soil, gardens, pastures and woodlands, and found to be abundant in soils bordering rivers and lakes. It has also been sighted in botanical gardens, lawns, peat bog, compost and under manure

Figure 4.14 Dark red, red or red–violet coloured earthworms (adapted from Wormwatch [12]).

Bimastos parvus

Common name: American bark worm, *Length*: 17–65 mm

Ventral view

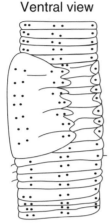

Habitat: This species is found in close association with habitats high in organic matter, such as decaying logs and leaves. Some species have also been found in moist areas associated with gardens, fields, manure and garbage dumps

Interesting facts and features: Its clitellum is found very close to its nose (< 1 cm)

Dendrobaena octaedra

Common name: Octagonal-tail worm, *Length*: 17–60 mm

Ventral view

Habitat: This species is commonly found in undistributed sites, such as under sod, moss on streams, logs and leafy debris

Interesting facts and features: It has widely paired setae

Figure 4.14 (*continued*).

Dendrodrilus rubidus
Common name: European bark worm, *Length*: 20–90 mm

Ventral view

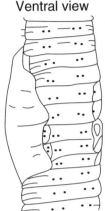

Habitat: This species has a broad range of habitats, including gardens, cultivated fields, stream banks, moss in running water, wells, springs, peat, compost and occasionally in manure. It can also be found under the bark of old rotting trees

Eisenia foetida
Common name: Manure worm, Tiger worm, Red Wiggler, *Length*: 30–60 mm

Ventral view

Habitat: This species is commonly found in very moist manure and organic matter. Examples of such environments include compost heaps and cowpats. Other locations are forests, gardens, and under stones, leaves, logs and roadside dumps

Interesting facts and features: This species in usually found in close proximity to human habitation and has been used for home composting and fish bait

Figure 4.14 (*continued*).

Lumbricus castaneus
Common name: Chestnut worm, *Length*: 30–50 mm (generally, < 35 mm)

Ventral view

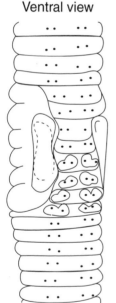

Habitat: This species is terrestrial and is commonly found in soil rich in organic matter, such as gardens, parks, pastures, forests, riverbanks, marsh banks or under stones, leaves and dung. It can also be found in cultivated fields, manure, compost and leaf litter

Lumbricus festivus
Common name: Quebec worm, *Length*: 48–105 mm

Ventral view

Habitat: Little is known about this rare species' habitat since few have been found. Most of these earthworms have been found under logs, debris, dung, stones and leaf piles, as well as in pastures and riverbanks

Interesting facts and features: A very rare earthworm

Figure 4.14 (*continued*).

Lumbricus rubellus
Common name: Red marsh worm, *Length*: 50–150 mm (generally, > 60 mm)

Ventral view

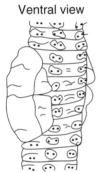

Habitat: This species is commonly found in places rich in organic matter and moisture, such as parks, gardens, pastures, woody peat, riverbanks and under stones, moss, old leaves, dung in pastures and logs

Interesting facts and features: This species has been cultured by the fish-bait industry

Lumbricus terrestris
Common name: Nightcrawler, dew worm, *Length*: 90–300 mm

Ventral view

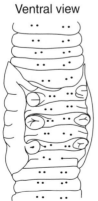

Habitat: This species is almost purely terrestrial. It is commonly found in gardens, lawns, pastureland and under logs. It is also found in forests, riverbanks, streams, mud flats, woody peat, under cowpats and in compost

Interesting facts and features: It is very long-lived (up to 10 years) and it forms deep vertical burrows

Figure 4.14 (*continued*).

Allolobophora chlorotica
Common name: Green worm, manure worm, *Length*: 30–70 mm

Ventral view

Habitat: This species is found in wet areas. It has also been found in gardens, fields, pastures, forests, estuary flats, lake shores and in manure

Interesting facts and features: This species is easily recognized by its muddy green colour and sucker-like tubercula pubertatis (TP)

Figure 4.15 Muddy green coloured earthworms (adapted from Wormwatch [13]).

Eiseniella tetraedra
Common name: Square tail worm, *Length*: 30–60 mm

Ventral view

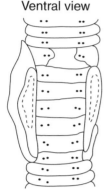

Habitat: This species is commonly found in damp habitats, such as wells, springs, moss of swift streams, subterranean waters, rivers ponds, lakes, canals and water-soaked banks of streams

Figure 4.15 (*continued*).

Aporrectodea icteria
Common name: Mottled worm, *Length*: 55–135 mm

Ventral view

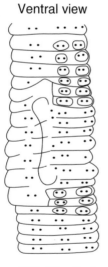

Habitat: This species has been found in garden soil, meadows and orchards

Figure 4.16 Other body colours of earthworms (adapted from Wormwatch [14]).

Aporrectodea rosea

Common name: Pink soil worm, *Length*: 25–85 mm

Ventral view

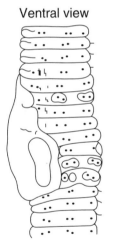

Habitat: This species is commonly found in fields, gardens, pastures, forests and under leaves and stones. It has also been spotted along the shores of rivers and lakes

Interesting facts and features: This species has a unique flared clitellum, a pink head and a grey body

Aporrectodea trapezoides

Common name: Southern worm, *Length*: 80–140 mm

Ventral view

Habitat: This species is commonly found in the soil around the roots of potted plants, in gardens, cultivated fields, forests, soils of various types, on the banks of streams and occasionally in sandy soils

Interesting facts and features: This earthworm is reasonably 'drought-tolerant'

Figure 4.16 (*continued*).

Aporrectodea tuberculata

Common name: Canadian worm, *Length*: 90–150 mm

Ventral view

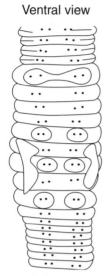

Habitat: This species is widely distributed in most habitats. It has been found in wet areas near streams and springs where there is a large concentration of organic matter. Other sightings include under logs, compost, peat, rocks, ditches, turf and occasionally in manure

Aporrectodea turgida

Common name: Pasture worm, *Length*: 60–85 mm

Ventral view

Habitat: This species has a broad range of habitats, including gardens, fields, turf, leaf litter in forests, compost, banks of spring-sand streams, wasteland, city dumps and streams. It is commonly found in irrigated areas

Figure 4.16 (*continued*).

Octolasion cyaneum
Common name: Woodland blue worm, *Length*: 65–180 mm

Ventral view

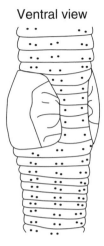

Habitat: This species has been found under stones, in water, in moss, stream banks, ploughed fields, wet sand and forest soil. It is also commonly found under logs and rocks near streambeds

Interesting facts and features: This species can be recognized by its bluish hue

Octolasion tyrtaeum
Common name: Woodland white worm, *Length*: 25–130 mm

Ventral view

Habitat: This species is commonly found under stones, logs, peat, leaf mold, compost, forest litter, gardens, cultivated fields, pastures, stream banks, in springs and around the roots of submerged vegetation

Interesting facts and features: This species can be recognized by its snub nose, light colour (almost grey) and the long distance between the clitellum and the nose (> 2 cm)

Figure 4.16 (*continued*).

Sparganophilus eiseni
Common name: American mud worm, *Length*: 150–200 mm

Ventral view

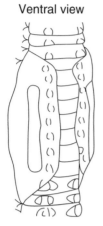

Habitat: This species is often found living in muddy areas. Examples of such locations include the muddy banks of streams, rivers, ponds and lakes

Interesting facts and features: Its clitellum is located unusually close to its nose (< 1 cm)

Figure 4.16 (*continued*).

- Earthworms reside in contaminated sites, allowing field validation of chemical bioavailability.

- Earthworms are found in a variety of soil types and horizons.

- The exterior epidermal surface of the earthworm is vascularized with no cuticle, hence allowing the uptake of contaminants directly from the soil.

- Earthworms ingest soil or specific fractions of soil, so providing a means for the dietary uptake of contaminants.

- Earthworms have a large mass and so contaminant concentrations can be determined in individual organisms.

- There is a low level of mixed-function oxidase (MFO) activity, hence allowing greater potential for the accumulation of organic compounds that would normally be metabolized in other organisms.

- An understanding exists of their physiology and metabolism of metals.

- Standardized toxicity testing protocols are available.

- Many techniques are available for assessing effects at population, organismal and sub-organismal levels for earthworms.

- Some species, such as *Eisenia fetida*, can be cultured in the laboratory under controlled conditions and are tolerant of many soil types, thus allowing the testing of different soils.

Table 4.7 Major earthworm protocols and studies (adapted from Sandoval *et al.* [16])

Protocol	Description	Comments
ASTM 1676–95: *Standard Guide for Conducting Laboratory Soil Toxicity Tests for the Lumbricid Earthworm* Eisenia foetida [17]	Using the earthworm *Eisenia foetida*, a 28-day test was devised; this involved the creation of a sphagnum-based synthetic soil to be mixed with the test media	• Acute toxicity test • Complexation of synthetic soil/organics with metals possible; also variability of synthetic soils from 'test-to-test' likely
US EPA 600/R-94/024: *Methods for Measuring the Toxicity and Bioaccumulation of Sediment-Associated Contaminants with Freshwater* Invertebrates [18] ASTM 1383–93a: *Standard Guide for Conducting Sediment Toxicity Tests with Freshwater Invertebrates* [19]	Using the aquatic invertebrate *Lumbriculus variegates,* a 28-day test for sediments was devised; this involved a fully saturated system and continuous 'flow-through' of overlying water	• Limited to sediments • Requires collection and treatment of water
Lockheed Martin Environmental Restoration Program: *Development and Validation of Bioaccumulation Models for Earthworms* [20]	Compared bioaccumulation of selected organics and inorganics by earthworms using 33 studies from 12 countries and developed bioaccumulation models/numerical correlations	• Does not provide specific methodology information on studies reviewed • Numerical correlation between earthworms and substrates derived from various protocols
Goats and Edwards: *The Prediction of Field Toxicity of Chemicals to Earthworms by Laboratory Methods* [21]	Review of available methods for toxicity prediction using earthworms and development of new methods; evaluated various synthetic and natural substrates, modes of exposure and compared field methods for toxicity to the same series of contaminants	• Evaluated contaminants in solutions only • Acute toxicity test • Standardized substrate (using filter paper or silica) deemed most appropriate for solution evaluation

(*continued overleaf*)

Table 4.7 (*continued*)

Protocol	Description	Comments
Callahan: *Earthworms as Ecotoxicological Assessment Tools (Hazardous Materials Assessment Team (HMAT) 14-Day Soil Test Using Earthworms)* [22]	Involves addition of contaminant of concern with standardized mixture of sand, clay and peat and exposure of worms for 14 days	• Evaluated contaminants in solutions only • Bioaccumulation test • Contaminant interactions with prepared soil probable

Earthworms protocols have been reviewed by Sandoval *et al.* [16] and are shown in Table 4.7. As a result of their work, they proposed a new earthworm protocol for metal analysis.

4.4.1.1 Proposed Earthworm Protocol[†]

Specific components of the methodology, including exposure time, the composition of the substrate, physiological requirements of the earthworms (pH, humidity, organic carbon) and procedures to void contents of the intestinal tract (so that only tissue is analysed) were derived from a review of these major studies and other laboratory and field-scale studies. The methodology is summarized as follows and described in greater detail below:

- A mixture of 80 g of tailings, sediments or clean sand (for solution assessments), 20 g of prepared cellulose and 80 ml of distilled water (in solids evaluation) or solution are manually homogenized in a 900 ml glass jar (45% moisture content).

- Fifteen cleaned, weighed worms (*Eisenia foetida*) are placed in the mixture for a period of 28 days.

- At the conclusion of the exposure period, the worms are removed, counted, cleaned and depurated for 24 h to void contents of the stomach and intestinal tract.

- Depurated worms are re-cleaned and weighed and pre-digested for 'full-metals' scanning.

- Jar contents are also sampled and analysed for metals for comparison with worm tissues.

The details of the methodology are described in the following.

[†] From Sandoval *et al.* [16].

Organism Culturing and Selection　*E. foetida* can be initially acquired from local composting co-operatives and cultured in a dark plastic, ventilated bin. Diets of alfalfa pellets, a mixture of vegetable food waste or horse manure, have been compared and it was determined that optimum growth occurs in a pure horse manure substrate. Worms are hand-selected for testing on the basis of sexual maturity, as evidenced by the presence of a clitellum (a 3 mm-wide ring around the body), size (0.3–0.45 g wet weight) and liveliness (actively responds when anterior segments are prodded). Prior to use in jar experiments, chosen worms are stored for 24 h on damp filter paper to void contents of the stomach and intestinal tract.

Cellulose Preparation　Cellulose is prepared in advance by shredding white, kaolin-based paper, followed by converting to a pulp by mixing in a blender with distilled water and subsequently drying at 30°C for 48 h. Dried paper can be broken down into a softer cellulose mixture by using a blender.

Tailings/Soil/Sediment Jar Preparation　Tailings or contaminated soil (80 g) are combined with cellulose (20 g) and manually homogenized with distilled water (80 ml) in 900 ml jars. Fifteen mature earthworms are placed in each jar (in triplicate) for a period of 28 days (deemed to be 'steady-state'). A blank containing cellulose, water and clean sand is also prepared in conjunction with the test jars.

Solutions/Effluents Jar Preparation　Clean, fine sand (80 g) is combined with cellulose (20 g) and manually homogenized with solution to be evaluated (80 ml) in acid-washed 900 ml jars. Fifteen mature earthworms are placed in each jar (in triplicate) for a period of 28 days. A blank containing cellulose, water and clean sand is also prepared in conjunction with the test jars. The proportions used in the protocol were derived empirically from previous tests conducted by these researchers, previous studies and knowledge of earthworms' natural habitat. Earthworms thrive in moisture contents between 30–45% and at a pH ranging from 5 to 9. For the evaluation of effluents, some pH adjustment (ideally using NaOH) to achieve an amenable pH may be required.

Post-Exposure Period　At the conclusion of the exposure period, worms are removed from each jar, carefully washed, dried and counted. Observations, such as motility, light sensitivity and physical qualities (e.g. discolouration), are documented to provide some indication of toxic responses. Following this, the worms are depurated (i.e. 'starved') for a period of 24 h to void contents of the intestinal tract and subsequently re-washed and re-weighed. Both worms and jar contents are analysed for 'total metals' to determine bioaccumulation following worm digestion.

Worm Digestion and Analysis   Following post-depuration washing and drying, worms are placed in 250 ml, acid-washed Erlenmeyer flasks and digested in 20 ml of 0.7 M nitric acid. After a 24-h period, the solution is slowly reduced to 10 ml at low heat. Distilled water is then added up to a volume of 120 ml. Samples are split into 60 ml volumes, poured into acid-washed polyethylene containers and promptly refrigerated. One of the two samples is kept as a 'back-up' and the other submitted for full-metal analysis (e.g. by ICP–MS).

4.4.2 Chemical-Extraction Methods to Estimate Bioavailability of Metals by Earthworms

Previous work [23] has indicated that the use of single/sequential extraction methods give data, which although not perfect, produce an improved approach for evaluating soil metal toxicity when compared to total metal concentrations. One approach that has been used is the use of DTPA extraction.

SAQ 4.4

What is the chemical name of DTPA?

As discussed previously (Section 4.2), DTPA provides a single extraction approach that is widely used to assess metal mobility in soils. DTPA is capable of releasing metals that are soluble, exchangeable, adsorbed, organically bound and possibly some that are fixed to oxides [24].

Dai *et al.* [25] investigated the link between heavy metal accumulation (Cd, Cu, Pb and Zn) by two earthworm species (*Aporrectodea caliginosa and Lumbricus rubellus*) and its relationship to total and DTPA-extractable metals in soils. Soil samples were acquired from a site that had been historically contaminated with heavy metals. Soil samples and worms were collected from the six different locations across the site known to have increasing levels of metal contamination. The total metal and DTPA-extracted metal data are shown in Figure 4.17.

DQ 4.11

Comment on the results shown in this figure.

Answer

The amount of metal in the DTPA extracts, compared to total metal, never exceeded 49% for Cd, 33–34% for Zn and Pb, and 22% for Cu.

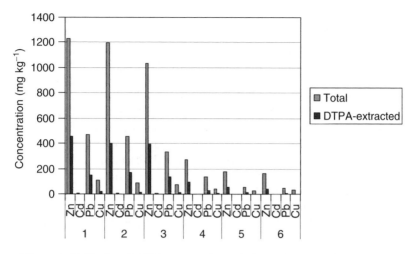

Figure 4.17 Total and DTPA-extracted metals from six sampling sites [25].

In contrast, the metal content of the earthworms on the different sites ranged from 11.6 to 102.9 mg kg^{-1} for Cd, 17.9 to 35.9 mg kg^{-1} for Cu, 1.9 to 182.8 mg kg^{-1} for Pb and 556 to 3381 mg kg^{-1} for Zn in *Aporrectodea caliginosa*, while in *Lumbricus rubellus* the concentrations were 7.7 to 26.3 mg kg^{-1} for Cd, 16.0

Table 4.8 Bioconcentration of heavy metals in earthworms (*Lumbricus rubellus* and *Aporrectodea caliginosa*) and their correlations with total metal and DTPA-extracted metal contents in soil samples [25]

Metal	Content	*Lumbricus rubellus*		*Aporrectodea caliginosa*	
		Correlation coefficient, r	Probability, P	Correlation coefficient, r	Probability, P
Cd	Total	0.36	*0.15*	0.49	0.04
	DTPA-extracted	0.35	*0.16*	0.48	0.05
Cu	Total	0.44	0.07	0.42	*0.08*
	DTPA-extracted	0.54	0.02	0.42	*0.09*
Pb	Total	0.62	0.007	0.47	0.05
	DTPA-extracted	0.73	0.0005	0.57	0.01
Zn	Total	0.59	0.01	0.74	0.0004
	DTPA-extracted	0.62	0.006	0.76	0.0003

to 37.6 mg kg^{-1} for Cu, 0.5 to 37.9 mg kg^{-1} for Pb and 667.9 to 2645 mg kg^{-1} for Zn.

Statistical analysis, using linear correlations, between the metal concentrations in the earthworms and the total metal and DTPA-extracted metal contents of soil showed significant correlations (Table 4.8). Exemptions to this were the data for Cd in *Lumbricus rubellus* and Cu in *Aporrectodea caliginosa* (indicated by italics in the table). The authors indicated that DTPA-extracted metals may be a preferable predictor of the bioconcentrations of heavy metals in earthworms.

Alternative approaches [26, 27] have evaluated the use of a weak electrolyte, i.e. 0.1 M Ca(NO$_3$)$_2$, as an indicator of metal (Cd, Pb and Zn) bioavailability in soil and compared the data to lethal toxicity in the earthworm *E. fetida*. They concluded that Ca(NO$_3$)$_2$ showed promise as a precise and inexpensive approach to assess metal bioavailability in soil.

It is clear that while significant research has been devoted to establishing and correlating links between metal availability in soil, using single/sequential extraction approaches and the bioavailability of metal uptake by earthworms, more work is required.

4.5 Plant Uptake

Uptake by plants from soil is being investigated from a number of viewpoints, which include phytoremediation [28] and bioavailability:

- *Phytoremediation*

  – *Phytoextraction or phytoaccumulation.* Use of plants to accumulate contaminants into the roots and above ground shoots or leaves.

  – *Phytodegradation or phytotransformation.* Refers to the uptake of contaminants that are then transformed to more stable, less-toxic or less-mobile species (more likely to occur for organic pollutants, although Cr(VI) can be reduced to Cr(III) which is less mobile and non-carcinogenic).

  – *Phytostabilization.* Use of plants to lessen mobility and bioavailability of metals in contaminated soil by, for example, stabilizing the soil surface by root structure and/or vegetation cover.

  – *Rhizodegradation.* Breakdown of contaminants through activity in the rhizosphere (roots).

  – *Rhizofiltration.* A water-remediation technique involving uptake of contaminants by plant roots.

- *Bioavailability*

  - *Plant bioassays.* Establishment of relationships between metal plant uptake and either chemical-extraction approaches or uptake by earthworms.

While phytoremediation is outside the scope of this present book, the principles involved in plant uptake by metals are similar. A number of recent articles highlight research that is being undertaken on the uptake of metals by plants for phytoremediation purposes. A summary of these is given in Table 4.9.

The use of plants as bioassays is an important measure of bioavailability for metals (and persistent organic pollutants).

DQ 4.12

Can you think of two important aspects of this type of research?

Answer

The two aspects of this research are as follows:

- Analysis of plant tissue to assess whether levels of contamination within the soil are at elevated or potentially toxic levels.

- Determination of the growth and vigour of plants on contaminated soil. If the plant grows, then it is possible to conclude that the level of contamination is not at a phytotoxic concentration.

Both bioassays are used to assess the bioavailability of contaminants to plants and also to animals that consume the plants. In addition, these approaches provide an estimate of the bioavailability of contaminants to other organisms, providing that a correlation can be established between plant and animal uptake, e.g. earthworms studies (see Section 4.4). A number of recent articles highlight research that is being undertaken on the uptake of metals by plants for bioavailability purposes. A summary is given in Table 4.9.

4.6 Certified Reference Materials

A Certified Reference Material (CRM) is a substance for which one or more analytes (metals or POPs) have certified values, produced by a technically valid procedure, accompanied with a traceable certificate and issued by a certifying body.

Table 4.9 Examples of recent research investigating the uptake of metals by plants[a]
(a) *Phytoremediation*

Metal	Plant type	Comments	Extraction/Analysis	Reference
As	Indigenous plant species (36) from Thailand	On the basis of high As tolerance, high bioaccumulation factor, short-life cycle, high propagation rate, wide distribution and large shoot biomass, the following plants were selected as useful for phytoremediation: two ferns (*Pityrogramma calomelanos* and *Pteris vittata*), a herb (*Mimosa pudica*) and a shrub (*Melastoma malabrathricum*)	HyAAS samples were dry-ashed, while GFAAS samples were acid-digested using 6 M HCl and HCl:HNO$_3$ (3:1, vol/vol). Analysis by HyAAS and GFAAS	Visoottiviseth *et al.* [29]
Pb	Alfalfa (*Medicago sativa*)	Enhanced uptake of metal using EDTA and a plant growth promoter (indole-3-acetic acid) for phytoremediation. Plants were grown hydroponically	Samples digested using microwave digestion with HNO$_3$. Analysis by ICP–AES	Lopez *et al.* [30]
Pb	Wheat (*Triticum aestivum L.*)	Chelate-assisted (EDTA and acetic acid) phytoextraction of metal from soils by plants. Enhancement of shoot Pb uptake when both EDTA and acetic acid were applied one week before harvesting	Samples digested with HNO$_3$ and H$_2$O$_2$. Analysis by AAS	Begonia *et al.* [31]

Metals	Plant	Description	Method	Reference
As, Cd, Pb and Zn	Lettuce, pumpkin, zucchini, cucumber, tomato, thistle, lupin and spinach	An approach to phytoremediating metals and POPs simultaneously	Samples digested with concentrated HNO_3 on a hot-plate. Analysis by ICP–AES	Mattina *et al.* [32]
As, Cd, Pb and Zn	Grass, lupine and yellow mustard	Chelate-enhanced (EDGA and citric acid) phytoremediation for extraction of metals from soils by plants. Citric acid was microbially degraded, whereas EDGA-enhanced metal solubility did not increase plant uptake	Samples digested with concentrated HNO_3. Analysis by ICP–AES	Romkens *et al.* [33]
Cd, Cr, Cu, Ni, Pb and Zn	Canola (*Brassica napus*) and radish (*Raphanus sativus*)	Phytoextraction of heavy metals from polluted soils. Both plants showed relatively low phytoremediation potential of multicontaminated soils	Samples digested using microwave-assisted total digestion. Analysis by ICP–AES	Marchiol *et al.* [34]
Cd, Cr, Cu, Ni, Pb and Zn	*Helianthus annus*	Phytoextraction of heavy metals from soils treated with EDTA, EDDS, NTA and citric acid. Uptake of heavy metals was greatest in EDDS-treated soils compared to EDTA-treated soils. No effects noted for NTA and citric acid	Samples digested with aqua regia. Analysis by ICP–AES	Meers *et al.* [35]

(continued overleaf)

Table 4.9 (*continued*)

Metal	Plant type	Comments	Extraction/Analysis	Reference
Cd, Cu, Ni, Pb and Zn	—	Phytoextraction of heavy metals from exogenous humic acid-treated soils. It was concluded that the humic acid-treated soil may accelerate phytoremediation of heavy metals from contaminated soil, while concomitantly preventing the environmental mobility of the metals	Samples extracted using (i) 2.5% glacial acetic acid for readily soluble and easily exchangeable forms of metals, (ii) 0.005 M DTPA in 0.01 N $CaCl_2$ and 0.1 M triethenolamine at pH 7.3 for 'plant-available' metals and (iii) 1 M HNO_3 for metals occluded in carbonates, phosphates and oxides. Analysis by FAAS	Halim *et al.* [36]
Cd and Zn	*Thlaspi caerulescens* and *Arabidopsis halleri*	Field trials to evaluate effectiveness of metal-mobilizing agents (EDTA, NTA and citric acid), different sowing strategies and length of growing season on phytoextraction of metals by plants. *T. caerulescens* could be used to 'clean-up' soil moderately contaminated with Cd	Samples digested with HNO_3 and $HClO_4$. Analysis by ICP–AES	McGrath *et al.* [37]

| Pb and Zn | *Lolium italicum* and *Testuca arundinacea* | Phytostabilization of a site downstream of a former mining area in Italy. No clear effects noted | Samples digested with $HNO_3/HClO_4$ (2.5:1, vol/vol) in a microwave oven for total metal and soil-extractable metals using sequential extraction with H_2O, 1 M KNO_3 and 0.01 M EDTA. Analysis by FAAS | Rizzi *et al.* [38] |

(b) *Bioavailability*

Metal	Plant type	Comments	Extraction/Analysis	Reference
Cu	Tomato	Bioavailability and extractability as related to the chemical properties of contaminated soil in the vine-growing area investigated. Simple extractions routinely used in soil-testing procedures (total and EDTA-extractable) were adequate indicators of bioavailability for calcareous soils but not for other soils considered, e.g. acidic soils	Samples digested with $HF/HClO_4$. Analysis by FAAS	Chaignon *et al.* [39]

(continued overleaf)

Table 4.9 (*continued*)

Metal	Plant type	Comments	Extraction/Analysis	Reference
Pb	Tea leaves	High levels of metal in tea leaves investigated. 47% of samples investigated exceeded the permissible level in China ($2\,mg\,kg^{-1}$ Pb). It was reported that soil pH and organic matter accumulation contributed to an increase in Pb uptake and accumulation in tea leaves	Samples digested with HCl/HClO$_3$ (4:1, vol/vol). Analysis by GFAAS	Jin *et al.* [40]
As, Cd, Cr, Cu, Hg, Mn, Ni, Pb and Zn	Wheat	Soil-to-plant transfer factors (TFs) determined for a range of metals. Extraction procedure evaluated to determine 'plant-available' metal. Several extraction procedures (12) evaluated but the best was 2% (wt/vol) ammonium citrate	Samples digested with HNO$_3$ in a microwave oven. Analysis by ICP–AES	Chojnacka *et al.* [41]
Cd, Cu, Mn, Pb and Zn	Lettuce, spinach, radish and carrot	Soil-to-plant transfer factor (TF) values decreased as Mn $\gg$ Zn $>$ Cd $>$ Cu $>$ Pb. Individual plant types greatly differ in their metal uptake, e.g. spinach accumulated a high content of Mn and Zn, while relatively lower concentrations were found for Cu and Pb in their tissues	Samples digested with HNO$_3$/H$_2$SO$_4$/H$_2$O$_2$ on a hot-plate. Analysis by FAAS or DPASV	Intawongse and Dean [42]

Cu, Fe, Hg, K Nettle (*Urtica dioica*) and Mn	Metals sequentially extracted from soil using DTPA and 10% HNO_3. Correlations between the biological absorption coefficients (BACs) and mobile element absorption coefficients (MACs) with soil-extractable elements determined. DTPA extractions correlated best with aerial plant uptake of Cu, while Hg correlated best with plant-root BACs and MACs. Data compared with a bioassay using an earthworm (*E. fetida*)	Samples digested with H_2O_2/HNO_3 in a microwave oven. Analysis by FAAS for Cu, Fe and Mn, FAES for K and CV-AAS or AFS for Mg	Edwards *et al.* [43]

[a] AAS, atomic absorption spectroscopy; AFS, atomic fluorescence spectroscopy; $CaCl_2$, calcium chloride; CV-AAS, cold-vapour atomic absorption spectroscopy; DPASV, differential-pulse anodic stripping voltammetry; DTPA, diethylenetriamine pentaacetic acid; EDDS, ethylenediamine disuccinate; EDGA, glycoletherdiamine tetraacetic acid; EDTA, ethylenediamine tetraacetic acid; FAAS, flame atomic absorption spectroscopy; FAES, flame atomic emission spectroscopy; GFAAS, graphite-furnace atomic absorption spectroscopy; $HClO_3$, chloric acid; $HClO_4$, perchloric acid; H_2O_2, hydrogen peroxide; HNO_3, nitric acid; HyAAS, hydride-generation atomic absorption spectroscopy; ICP–AES, inductively coupled plasma–atomic emission spectroscopy; KNO_3, potassium nitrate; NTA, nitrilo triacetic acid.

Table 4.10 Certified Reference Materials for single and sequential extractable elements from soil and sediment matrices [44]

CRM code	Matrix	Metals	Concentration range	Comments
BCR701	Sediment	Cd, Cr, Cu, Ni, Pb and Zn	2.26–205 mg kg^{-1}; 3.77–126 mg kg^{-1}; 0.27–143 mg kg^{-1}	Three-step sequential extraction (Step 1; Step 2; Step 3)
BCR684	River sediment	P	550 mg kg^{-1}; 536 mg kg^{-1}; 1113 mg kg^{-1}; 209 mg kg^{-1}; 1373 mg kg^{-1}	NaOH-extractable; HCl-extractable; inorganic; organic; concentrated HCl-extractable
BCR700	Organic-rich soil	Cd, Cr, Cu, Ni, Pb and Zn	10.1–510 mg kg^{-1}	EDTA-extractable
	Organic-rich soil	Cd, Cr, Cu, Ni, Pb and Zn	4.85–719 mg kg^{-1}	Acetic acid-extractable
RTH907–RTH972	Soils	Ring-test programme operated by Wageningen Agricultural University, The Netherlands		0.1M NaNO$_3$; 0.01M CaCl$_2$; 1M NH$_4$ NO$_3$; 1M NH$_4$ acetate; BaCl$_2$
BCR483	Sewage 'sludge-amended' soil	Cd, Cr, Cu, Ni, Pb and Zn	24.3–612 mg kg^{-1}	EDTA-extractable
			2.10–620 mg kg^{-1}	Acetic acid-extractable
BCR484	Sewage 'sludge-amended' soil	Cd, Cu, Ni, Pb and Zn	0.51–383 mg kg^{-1}	EDTA-extractable
			0.48–193 mg kg^{-1}	Acetic acid-extractable

Table 4.11 Certified Reference Materials for acid-extractable metals from soil and sediment matrices [44]

CRM code	Matrix	Metals	Concentration range[a]	Comments
LGC6139	Freshwater sediment	As, Cd, Cr, Cu, Hg, Ni, Pb, Se and Zn	$0.59-513\,\mathrm{mg\,kg^{-1}}$	Hot aqua regia-extractable
LGC6187	River sediment	As, Cd, Cr, Cu, Fe, Hg, Mn, Ni, Pb, Se, Sn, V and Zn	$1.2-23\,600\,\mathrm{mg\,kg^{-1}}$	Hot mixture of HNO_3 and HCl-extractable
CAN-LKSD 1	Sediment	Ag, As, Cd, Co, Cr, Cu, Fe, Hg, Mn, Mo, Ni, Pb, Sb, V and Zn	$0.6-460\,\mathrm{mg\,kg^{-1}}^{a}$	Aqua regia-extractable
	Sediment	Ag, Cd, Co, Cu, Fe, Mn, Ni, Pb and Zn	$0.6-410\,\mathrm{mg\,kg^{-1}}^{a}$	Dilute aqua regia-extractable
CAN-LKSD 2	Sediment	Ag, As, Cd, Co, Cr, Cu, Fe, Hg, Mn, Mo, Ni, Pb, Sb, V and Zn	$0.8-1840\,\mathrm{mg\,kg^{-1}}^{a}$	Aqua regia-extractable
	Sediment	Ag, Cd, Co, Cu, Fe, Mn, Ni, Pb and Zn	$0.6-1840\,\mathrm{mg\,kg^{-1}}^{a}$	Dilute aqua regia-extractable
CAN-LKSD 3	Sediment	Ag, As, Cd, Co, Cr, Cu, Fe, Hg, Mn, Mo, Ni, Pb, Sb, V and Zn	$0.6-1220\,\mathrm{mg\,kg^{-1}}^{a}$	Aqua regia-extractable
	Sediment	Ag, Cd, Co, Cu, Fe, Mn, Ni, Pb and Zn	$0.4-1300\,\mathrm{mg\,kg^{-1}}^{a}$	Dilute aqua regia-extractable

(continued overleaf)

Table 4.11 (*continued*)

CRM code	Matrix	Metals	Concentration range[a]	Comments
CAN-LKSD 4	Sediment	Ag, As, Cd, Co, Cr, Cu, Fe, Hg, Mn, Mo, Ni, Pb, Sb, V and Zn	$0.2–430\,\mathrm{mg\,kg^{-1}}$[a]	Aqua regia-extractable
	Sediment	Ag, Cd, Co, Cu, Fe, Mn, Ni, Pb and Zn	$0.2–420\,\mathrm{mg\,kg^{-1}}$[a]	Dilute aqua regia-extractable
CAN-STSD 1	Sediment	Ag, As, Cd, Co, Cr, Cu, Fe, Hg, Mn, Mo, Ni, Pb, Sb, V and Zn	$0.3–3740\,\mathrm{mg\,kg^{-1}}$[a]	Aqua regia-extractable
CAN-STSD 2	Sediment	Ag, As, Cd, Co, Cr, Cu, Fe, Hg, Mn, Mo, Ni, Pb, Sb, V and Zn	$0.5–720\,\mathrm{mg\,kg^{-1}}$[a]	Aqua regia-extractable
CAN-STSD 3	Sediment	Ag, As, Cd, Co, Cr, Cu, Fe, Hg, Mn, Mo, Ni, Pb, Sb, V and Zn	$0.4–2630\,\mathrm{mg\,kg^{-1}}$[a]	Aqua regia-extractable
CAN-STSD 4	Sediment	Ag, As, Cd, Co, Cr, Cu, Fe, Hg, Mn, Mo, Ni, Pb, Sb, V and Zn	$0.3–1200\,\mathrm{mg\,kg^{-1}}$[a]	Aqua regia-extractable
LGC6156	Harbour sediment	Al, As, Be, Ca, Cd, Co, Cr, Cu, Fe, Hg, K, Mg, Mn, Mo, Na, Ni, Pb, Sn, V and Zn	$2.9–20\,100\,\mathrm{mg\,kg^{-1}}$[a]	Aqua regia-extractable

Table 4.11 (*continued*)

CRM code	Matrix	Metals	Concentration range[a]	Comments
BCR142R	Light sandy soil	Cd, Ni, Pb and Zn	$0.25-93.3\,mg\,kg^{-1}$	Aqua regia-extractable
CMI7001	Light sandy soil	As, Ba, Be, Cd, Cu, Hg, Mn, Ni, Pb, V and Zn	$0.29-479\,mg\,kg^{-1}$	Aqua regia-extractable
	Light sandy soil	As, Be, Co, Cr, Cu, Mn, Ni, Pb, V and Zn	$0.71-438\,mg\,kg^{-1}$	Boiling $2\,mol\,l^{-1}$ HNO_3-extractable
	Light sandy soil	As, Be, Cd, Co, Cr, Cu, Mn, Ni, Pb, V and Zn	$0.18-357\,mg\,kg^{-1}$	Cold $2\,mol\,l^{-1}$ HNO_3-extractable
CMI7002	Light sandy soil	As, Be, Cd, Co, Cr, Cu, Mn, Ni, Pb, V and Zn	$0.28-531\,mg\,kg^{-1}$	Aqua regia-extractable
	Light sandy soil	As, Be, Cd, Co, Cr, Cu, Hg, Mn, Ni, Pb, V and Zn	$0.046-481\,mg\,kg^{-1}$	Boiling $2\,mol\,l^{-1}$ HNO_3-extractable
	Light sandy soil	As, Be, Cd, Co, Cr, Cu, Mn, Ni, Pb, V and Zn	$0.21-425\,mg\,kg^{-1}$	Cold $2\,mol\,l^{-1}$ HNO_3-extractable
BCR141R	Calcareous loam soil	Cd, Co, Cr, Cu, Hg, Mn, Ni, Pb and Zn	$0.24-653\,mg\,kg^{-1}$	Aqua regia-extractable
CMI7003	Silty clay loam soil	As, Be, Cd, Co, Cr, Cu, Mn, Ni, Pb, V and Zn	$0.32-529\,mg\,kg^{-1}$	Aqua regia-extractable
	Silty clay loam soil	Be, Cd, Co, Cr, Cu, Hg, Mn, Ni, Pb, V and Zn	$0.054-476\,mg\,kg^{-1}$	Boiling $2\,mol\,l^{-1}$ HNO_3-extractable

(*continued overleaf*)

Table 4.11 (*continued*)

CRM code	Matrix	Metals	Concentration range[a]	Comments
	Silty clay loam soil	As, Be, Cd, Co, Cr, Cu, Mn, Ni, Pb, V and Zn	0.23–435 mg kg^{-1}	Cold 2 mol l^{-1} HNO$_3$-extractable
CMI7004	Loam soil	As, Be, Cd, Co, Cr, Cu, Mn, Ni, Pb, V and Zn	1.44–741 mg kg^{-1}	Aqua regia-extractable
	Loam soil	As, Be, Cd, Co, Cr, Cu, Mn, Ni, Pb, V and Zn	1.44–572 mg kg^{-1}	Boiling 2 mol l^{-1} HNO$_3$-extractable
	Loam soil	As, Be, Cd, Co, Cr, Cu, Hg, Mn, Ni, Pb, V and Zn	0.094–527 mg kg^{-1}	Cold 2 mol l^{-1} HNO$_3$-extractable
RTH907-RTH972	Soils	Ring-test programme operated by Wageningen Agricultural University, The Netherlands		Boiling 2M HNO$_3$-extractable
ERML CC135a	Brickworks soil	Al, As, Ba, Be, Ca, Co, Cr, Cu, Fe, Hg, K, Mg, Mn, Na, Ni, Pb, Se, V and Zn	0.9–40 900 mg kg^{-1}	Aqua regia-extractable
LGC6141	Soil contaminated with clinker/ash	As, Cr, Cu, Ni, Pb and Zn	13.2–169 mg kg^{-1}	Aqua regia-extractable

Table 4.11 (*continued*)

CRM code	Matrix	Metals	Concentration range[a]	Comments
LGC6144	Gasworks-contaminated soil	As, Cr, Cu, Hg, Ni, Pb, Se, V and Zn	$0.53-192 \, mg \, kg^{-1}$	Aqua regia-extractable
BCR143R	Sewage 'sludge-amended' soil	Cd, Cr, Hg, Mn, Ni and Pb	$72-1063 \, mg \, kg^{-1}$	Aqua regia-extractable
ERMLCC136a	Sewage sludge	Al, Ba, Co, Cr, Cu, Fe, K, Mg, Mn, Na, Ni, Pb and Zn	$23.2-22\,200 \, mg \, kg^{-1}$	Aqua regia-extractable
LGC6181	Sewage sludge	Ag, As, Cd, Cu, Cr, Fe, Hg, Hg, Mn, Ni, Pb, V and Zn	$4.9-40\,300 \, mg \, kg^{-1}$	Aqua regia-extractable
BCR144R	Sewage sludge	Cd, Co, Cr, Cu, Hg, Mn, Ni, Pb and Zn	$1.84-919 \, mg \, kg^{-1}$	Aqua regia-extractable
BCR145R	Sewage sludge	Cr, Cu, Ni, Pb and Zn	$251-2137 \, mg \, kg^{-1}$	Aqua regia-extractable
BCR146R	Sewage sludge	Cd, Co, Cr, Cu, Hg, Mn, Ni, Pb and Zn	$6.5-3043 \, mg \, kg^{-1}$	Aqua regia-extractable

[a] Provisional.

DQ 4.13

Are you aware of any of these certifying bodies?

Answer

Common examples of certifying bodies are the National Institute for Standards and Technology (NIST), based in Washington, DC, USA, the Community Bureau of Reference (known as BCR), Brussels, Belgium and the Laboratory of the Government Chemist (LGC), London, UK.

CRMs (or Standard Reference Materials (SRMs)) can be either pure materials or matrix materials. Pure materials are used for the calibration of instruments, whereas matrix materials are used for the validation of a whole method from sample preparation through to the final measurement. It is the latter which is of interest in this section.

SAQ 4.5

Look at the Standard Reference Material (SRM) certificate shown in Figure 4.18 and comment on (a) the recording of numerical values and (b) the importance of the information given in footnote 2.

The relatively small number of CRMs [44] for the three-step BCR method and single extraction methods are shown in Table 4.10. Included within the table are data for the metals (and phosphorus) with certified values, an indication of the concentration range and the nature of the matrix. In addition to these CRMs for extractable metals using the approaches described in Sections 4.3.1. and 4.3.2, other CRMs are available that outline the acid (i.e. aqua regia and nitric acid)-extractable metals from soil and sediment matrices. Extraction with hot aqua regia is often described as providing metal recoveries that are referred to as the 'pseudo-total' (Table 4.11).

Summary

The different sample preparation techniques applied to the single and sequential extraction of metals from soils and sediments are described. In addition, the use of bioassays, in the form of earthworms and plants, for assessing metal bioavailability is reviewed. Finally, the role of Certified Reference Materials in the single and sequential extraction of metals from soils and sediments is considered.

National Institute of Science and Technology
Certificate of Analysis

Standard Reference Material 1515
Apple Leaves

Certified Concentrations of Constituent Elements[1]

Element	Concentration (wt%)
Calcium	1.526 ± 0.015
Magnesium	0.271 ± 0.008
Nitrogen (total)	2.25 ± 0.19
Phosphorus	0.159 ± 0.011
Potassium	1.61 ± 0.02

Element	Concentration (μg g^{-1})[2]	Element	Concentration (μg g^{-1})[2]
Aluminium	286 ± 9	Mercury	0.044 ± 0.004
Arsenic	0.038 ± 0.007	Molybdenum	0.094 ± 0.013
Barium	49 ± 2	Nickel	0.91 ± 0.12
Boron	27 ± 2	Rubidium	10.2 ± 1.5
Cadmium	0.013 ± 0.002	Selenium	0.050 ± 0.009
Chlorine	579 ± 23	Sodium	24.4 ± 12
Copper	5.64 ± 0.24	Strontium	25 ± 2
Iron	83 ± 5	Vanadium	0.26 ± 0.03
Lead	0.470 ± 0.024	Zinc	12.5 ± 0.3
Manganese	54 ± 3		

[1]The certified concentrations are equally weighted means of results from two or more different analytical methods or the means of results from a single method of known high accuracy.

[2]The values are based on dry weights. Samples of this SRM must be dried before weighing and analysis by, for example, drying in a desiccator at room temperature (ca. 22°C) for 120 h over fresh anhydrous magnesium perchlorate. The sample depth should not exceed 1 cm.

Figure 4.18 An example of a certificate of analysis for elements in apple leaves [45]. Reprinted from Certificate of Analysis, *Standard Reference Material 1515, Apple Leaves*, National Institute of Standards and Technology. Not copyrightable in the United States.

References

1. Ure, A. M., *Sci. Total Environ.*, **178**, 3–10 (1996).
2. Quevauviller, Ph., *Trends Anal. Chem.*, **17**, 289–298 (1998).
3. Quevauviller, Ph., *Trends Anal. Chem.*, **17**, 632–642 (1998).
4. Quevauviller, Ph., *Trends Anal. Chem.*, **21**, 774–785 (2002).
5. Pueyo, M., Lopez-Sanchez, J. F. and Rauret, G., *Anal. Chim. Acta*, **504**, 217–226 (2004).
6. Quevauviller, Ph., Rauret, G., Ure, A., Bacon, J. and Muntau, H., *The Certification of the EDTA- and Acetic Acid-Extractable Contents (Mass Fractions) of Cd, Cr, Cu, Ni, Pb and Zn in 'Sewage-Sludge'-Amended Soils*, Report EUR 17127 EN, European Commission, Brussels, Belgium, 1997.

7. Rauret, G., Lopez-Sanchez, J. F., Sahuquillo, A., Barahona, E., Lachica, M., Ure, A. M., Davidson, C. M., Gomez, A., Luck, D., Bacon, J., Yli-Halla, M., Muntau, H. and Quevauviller, Ph., *J. Environ. Monit.*, **2**, 228–239 (2000).

8. Mossop, K. F. and Davidson, C. M., *Anal. Chim. Acta*, **478**, 111–118 (2003).

9. Rauret, G., Lopez-Sanchez, J. F., Sahuquillo, A., Muntau, H. and Quevauviller, Ph., *Indicative Values for Extractable Contents (Mass Fractions) of Cd, Cr, Cu, Ni, Pb and Zn in Sediment (CRM 601) following the Modified BCR-Sequential Extraction (Three-Step) Procedure*, Report EUR 19502 EN, European Commission, Brussels, Belgium, 2000.

10. Rauret, G., Lopez-Sanchez, J. F., Sahuquillo, A., Barahona, E., Lachica, M., Ure, A., Muntau, H. and Quevauviller, Ph., *Indicative Values for Extractable Contents (Mass Fractions) of Cd, Cr, Cu, Ni, Pb and Zn in a Sewage sludge-Amended Soil (CRM 483) following the Modified BCR-Sequential Extraction (Three-Step) Procedure*, Report EUR 19503 EN, European Commission, Brussels, Belgium, 2000.

11. Wikipedia [http://en.wikipedia.org/wiki/Earthworm] (last accessed on 17 August, 2006).

12. Wormwatch [http://www.naturewatch.ca/english/wormwatch/about/guide/about_guide_redworms.html] (last accessed on 17 August, 2006).

13. Wormwatch [http://www.naturewatch.ca/english/wormwatch/about/guide/about_guide_greenworms.html] (last accessed on 17 August, 2006).

14. Wormwatch [http://www.naturewatch.ca/english/wormwatch/about/guide/about_guide_others.html] (last accessed on 17 August, 2006).

15. Lanno, R., Wells, J., Conder, J., Bradham, K. and Basta, N., *Ecotoxicol. Environ. Safety*, **57**, 39–47 (2004).

16. Sandoval, M. C., Veiga, M., Hinton, J. and Klein B., 'Review of biological indicators for metal mining effluents: a proposed protocol using earthworms,' in *Proceedings of the 25th Annual British Columbia Reclamation Symposium*, Campbell River, BC, Canada, September 23–27, British Columbia Technical and Research Committee on Reclamation (TRCR), Richmond, BC, Canada, 2001, pp. 67–79.

17. American Society for Testing and Materials (ASTM), *Standard Guide for Conducting Laboratory Soil Toxicity Tests for the Lumbricid Earthworm* Eisenia foetida, ASTM 1676–95, West Conshohocken, PA, USA, 1995.

18. United States Environmental Protection Agency (US EPA), *Methods for Measuring the Toxicity and Bioaccumulation of Sediment-Associated Contaminants with Freshwater Invertebrates*, EPA 600/R-94/024, Washington, DC, USA, 1994.

19. American Society for Testing and Materials (ASTM), *Standard Guide for Conducting Sediment Toxicity Tests with Freshwater Invertebrates*, ASTM 1383–93a, West Conshohocken, PA, USA, 1993.

20. Lockheed Martin Environmental Restoration Program, *Development and Validation of Bioaccumulation Models for Earthworms,* for the US Department of Energy, February, Lockheed Martin Environmental Restoration Program, Washington, DC, USA, 1998.

21. Goats, G. C. and Edwards, C. A., 'The prediction of field toxicity of chemicals to earthworms by laboratory methods, in *Earthworms in Waste and Environmental Management*, Edwards, C. A. and Neuhauser, E. F. (Eds), SPB. Academic Publishing, The Hague, The Netherlands, 1988, pp. 283–294.

22. Callahan, C. A., 'Earthworms as ecotoxicological assessment tools', in *Earthworms in Waste and Environmental Management*, Edwards, C. A. and Neuhauser, E. F. (Eds), SPB Academic Publishing, The Hague, The Netherlands, 1988, pp. 295–301.

23. McLaughlin, M. J., Zarcinas, B. A., Stevens, D. P. and Cook, N., *Commun. Soil Sci. Plant Anal.*, **31**, 1661–1700 (2000).

24. Lindsay, W. L. and Norvell, W. A., *Proc. Soil Soc. Am.*, **33**, 62–68 (1969).

25. Dai, J., Becquer, T., Rouiller, J. H., Reversat, G., Bernhard-Reversat, F., Nahmani, J. and Lavelle, P., *Soil Biol. Biochem.*, **36**, 91–98 (2004).

26. Conder, J. M. and Lanno, R. P., *Chemosphere*, **41**, 1659–1668 (2000).

27. Conder, J. M., Lanno, R. P. and Basta, N. T., *J. Environ. Qual.*, **30**, 1231–1237 (2001).

28. Vidali, M., *Pure Appl. Chem.*, **73**, 1163–1172 (2001).

29. Visoottiviseth, P., Francesconi, K. and Sridokchan, W., *Environ. Poll.*, **118**, 453–461 (2002).

30. Lopez, M. L., Peralta-Videa, J. R., Benitez, T. and Gardea-Torresdey, J. L., *Chemosphere*, **61**, 595–598 (2005).

31. Begonia, M. F. T., Begonia, G. B., Butler, A., Burrell, M., Oghoavodha, O. and Crudup, B., *Bull. Environ. Contam. Toxicol.*, **68**, 705–711 (2002).
32. Mattina, M. I., Lannucci-Berger, W., Musante, C. and White, J. C., *Environ. Poll.*, **124**, 375–378 (2003).
33. Romkens, P., Bouwman, L., Japenga, J. and Draaisma, C., *Environ. Poll.*, **116**, 109–121 (2002).
34. Marchiol, L., Assolari, S., Sacco, P. and Zerbi, G., *Environ. Poll.*, **132**, 21–27 (2004).
35. Meers, E., Ruttens, A., Hopgood, M. J., Samsom, D. and Tack, F. M. G., *Chemosphere*, **58**, 1011–1022 (2005).
36. Halim, M., Conte, P. and Piccolo, A., *Chemosphere*, **52**, 265–275 (2003).
37. McGrath, S. P., Lombi, E., Gray, C. W., Caille, N., Dunham., S. J. and Zhao, F. J., *Environ. Poll.*, **141**, 115–125 (2006).
38. Rizzi, L., Petruzzelli, G., Poggio, G. and Vigna Guidi, G., *Chemosphere*, **57**, 1039–1046 (2004).
39. Chaignon, V., Sanchez-Neira, I., Herrmann, P., Jaillard, B. and Hinsinger, P. *Environ. Poll.*, **123**, 229–238 (2003).
40. Jin, C. W., Zheng, S. J., He, Y. F., Zhou, G. D. and Zhou, Z. X., *Chemosphere*, **59**, 1151–1159 (2005).
41. Chojnacka, K., Chojnacki, A., Gorecka, H. and Gorecki, H., *Sci. Total Environ.*, **337**, 175–182 (2005).
42. Intawongse, M. and Dean, J. R., *Food Addit. Contam.*, **23**, 36–48 (2006).
43. Edwards, S. C., MacLeod, C. L. and Lester, J. N., *Water, Air Soil Poll.*, **102**, 75–90 (1998).
44. LGC-Promochem, *Reference Materials for Environmental Analysis, 2004/2005*, Laboratory of the Government Chemist, Teddington, UK [http://www.lgcpromochem.com/home/home_en.aspx] (last accessed on 8 January, 2006).
45. Dean, J. R., *Methods for Environmental Trace Analysis*, AnTS Series, John Wiley & Sons, Ltd, Chichester, UK, 2003.

Chapter 5

Methods Used to Assess Bioavailability of Persistent Organic Pollutants

---

Learning Objectives

- To appreciate the role that non-exhaustive extraction techniques have in the study of persistent organic pollutant bioavailability.
- To be aware of the use and procedures associated with a range of selective extraction methods for persistent organic pollutant bioavailability studies, including cyclodextrin, supercritical fluid extraction, sub-critical water extraction, solid-phase microextraction and membrane separations.
- To consider a mathematical approach for predicting solvent strength based on the Hildebrand solubility parameter.
- To appreciate the role that bioassays have in the study of persistent organic pollutant bioavailability.
- To be aware of the use and procedures associated with the use of earthworms for assessing persistent organic pollutant bioavailability.
- To be aware of the use and procedures associated with the use of plants for assessing persistent organic pollutant bioavailability.

5.1 Introduction

The recovery of the total amount of persistent organic pollutants (POPs) from environmental samples is important as an indication of the level of contamination.

Bioavailability, Bioaccessibility and Mobility of Environmental Contaminants   John R. Dean
© 2007 John Wiley & Sons, Ltd

However, while these total values can be used as predictors of environmental risk, they do not necessarily reflect the bioavailable fraction. Current techniques to recover POPs from environmental samples are often aggressive and designed to maximize recoveries.

SAQ 5.1

What techniques to recover POPs from environmental samples are you aware of?

An approach to assess the bioavailability of an analyte is to determine the amount recovered after aggressive sample preparation and to use this value as a measure of environmental toxicity. This approach will almost certainly lead to an 'over-approximation' of the environmental risk. In the case of POPs, this situation is more complex.

DQ 5.1

Why do you think that the situation with POPs may be more complex?

Answer

This is because the fate of POPs in the environment may be controlled by their susceptibility to chemical or biological degradation, as well as the possibility of retention by the soil matrix, i.e. they become 'sequestered'.

In the case of sequestration, this may be an effective solution for remediation as POPs that are sequestered are by definition inaccessible to micro-organisms, plants and animals. Therefore, sequestered POPs are not bioavailable and provide minimal environmental risk.

This chapter considers less aggressive sample preparation procedures which can lead to an understanding of the potential bioavailable fraction. In these approaches, it is normal to therefore expect that the recovered component will have a lower concentration than that obtained via the 'aggressive' procedures mentioned above which are designed to maximize recoveries. In this context, it is therefore important to use analytical techniques that are capable of detecting the lowest concentrations, so that the recovered components can be detected and quantified (see Chapter 3).

5.2 Non-Exhaustive Extraction Techniques for POPs

Non-exhaustive extraction techniques are often based on the use of organic solvents to selectively remove weakly bound POPs from solid matrices. This

Table 5.1 The properties of some common organic solvents [1]

Solvent	Structure	Boiling point (°C)	Dielectric constant (at 20°C)
Acetone	$H_3C-\overset{\overset{O}{\|\|}}{C}-CH_3$	56.1	20.7[a]
Acetonitrile	$H_3C-C\equiv N$	81.6	37.5
Carbon tetrachloride	CCl_4	76.8	2.238
Dichloromethane	CH_2Cl_2	39.8	9.08
Ethyl acetate	$H_3C-\overset{\overset{O}{\|\|}}{C}-OCH_2CH_3$	77.1	6.02[a]
Hexane	$CH_3(CH_2)_4CH_3$	68.7	1.890
Heptane	$CH_3(CH_2)_5CH_3$	98.4	1.92
Methanol	CH_3-OH	64.6	32.63[a]
Toluene	CH_3 (benzene ring)	110.6	2.379[a]

[a] At 25°C.

approach can be applied to any extraction technique which is used for the recovery of POPs from solid matrices. The important aspect is the choice of organic solvent to afford this selectivity. Table 5.1 shows the properties of some common solvents [1].

SAQ 5.2

How can organic solvents be classified?

The commonly accepted 'rule of thumb' is that polar organic compounds will dissolve in polar solvents.

DQ 5.2

On this basis, therefore, what would you expect to use to dissolve a non-polar organic compound?

Answer

A non-polar organic compound should therefore dissolve in a non-polar organic solvent, e.g. heptane.

In this current context, it is therefore important to assess how different organic solvents respond when used to recover POPs from environmental matrices. This section explores the role that the solvent has in the recovery of POPs from soil matrices (in the context of bioavailability studies) and the possibilities that exist to select the most appropriate solvent.

5.2.1 Selective or 'Mild-Solvent' Extraction

The sorption of organic pollutants to soils and sediments is considered to entail an initial rapid and reversible process, followed by a slow sorption process lasting for weeks, months or years (Figure 5.1). In this context, the initial sorption would be referred to as the 'bioavailable' fraction whereas the final process identifies the situation whereby the pollutants were resistant to desorption (the 'immobile' fraction). This resistance to recovery is often described as the 'aging' process. One approach to assess and identify the bioavailable fraction is to use 'mild' organic solvents to remove organic pollutants from soil. In this context, butan-1-ol has been extensively used as a mild organic solvent to recover organic pollutants from soils that have been aged (see for example, Hatzinger and Alexander [2], Kelsey *et al.* [3], Liste and Alexander [4] and Macleod and Semple [5]). Often this work has been linked to an assessment of the uptake of pollutants by earthworms (see Section 5.3) or bacterial degradation.

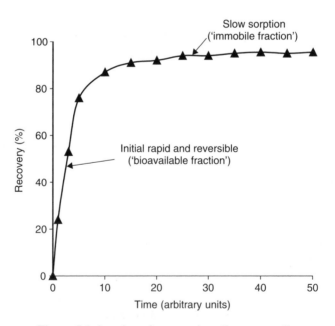

Figure 5.1 Sorption of an organic pollutant on soil.

The influence of aging (up to 84 days) on the recovery of ^{14}C-labelled phenanthrene[†] (and 4-nitrophenol) was reported by Hatzinger and Alexander [2]. They looked at the influence of three different soils (soil A, 19.3% organic matter and pH 6.9; soil B, 4.0% organic matter and pH 7.2; soil C, 2.3% organic matter and pH 7.4) on the recovery of phenanthrene by butan-1-ol. Extraction with this solvent consisted of mixing the soil (5 or 10 g) for 2 min with 20 or 25 ml of solvent, filtration and then measurement of ^{14}C-labelled phenanthrene in the supernatant by liquid scintillation counting. The results are shown in Table 5.2.

DQ 5.3

What do you observe about the results shown in Table 5.2?

Answer

It can be observed that soil A (highest soil organic matter) had the most significant effect on butan-1-ol recovery with aging, i.e. between 13 and 84 days the recovery of phenanthrene decreased to approximately

Table 5.2 Recovery of phenanthrene from soil by butan-1-ol after aging [2]

Soil identifier	Aging (days)	Amount extracted (%)[a]		
		Butan-1-ol	Soxhlet[b]	Total
Soil A	0	94.5 ± 3.9	9.0 ± 1.7	103.5 ± 3.6
	13	67.0 ± 7.2	24.8 ± 5.5	91.8 ± 1.8
	27	63.8 ± 6.2	24.5 ± 4.1	88.2 ± 2.2
	84	61.2 ± 5.5	25.9 ± 2.8	87.1 ± 3.7
Soil B	0	98.6 ± 1.8	6.1 ± 1.3	104.7 ± 3.1
	13	96.0 ± 3.5[c]	11.3 ± 0.3[c]	107.3 ± 3.8[c]
	27	88.8 ± 0.1[c]	10.7 ± 0.1[c]	99.5 ± 0.3[c]
	84	87.4 ± 3.1	12.7 ± 3.3	100.1 ± 1.3
Soil C	0	98.2 ± 0.9	4.2 ± 0.9	102.5 ± 1.1
	13	79.5 ± 2.1	22.0 ± 2.1	101.5 ± 0.6
	27	82.8 ± 9.0	20.4 ± 6.9	103.2 ± 3.1
	84	96.2 ± 4.5	5.2 ± 2.3	101.4 ± 2.8

[a]Mean ± standard deviation (SD) ($n = 3$).
[b]Mean ± standard deviation (SD) ($n = 2$).
[c]For 8 h using dichloromethane.

[†] The use of ^{14}C-labelled compounds, and subsequent measurement by liquid scintillation counting, allows an activity mass balance to be carried out, thus ensuring that all of the labelled compound can be accounted for.

60–70%. The effects were not as pronounced on soils B and C (i.e. 80–90% recovery after aging).

This approach was expanded to investigate a range of organic solvents to predict the bioavailability of phenanthrene (and atrazine) from soil [3] (Figure 5.2). The results were assessed alongside data for uptake (or bioaccumulation) by earthworms (*Eisenia foetida*) and bacterial mineralization (degradation of phenanthrene by the bacteria and subsequent evolution of labelled CO_2) by *Pseudomonas* strain R, for phenanthrene. The procedures for earthworms (Figure 5.3) and bacterial mineralization (Figure 5.4) studies are described. In the case of the former (Figure 5.3), the earthworms were added after defined aging periods had occurred, while in the case of the latter (Figure 5.4), $^{14}CO_2$ evolution to reflect bacterial mineralization by the *Pseudomonas* bacteria was measured after known time-periods had elapsed. The results are shown in Table 5.3 for the recovery of phenanthrene from aged soil over 7, 50 and 120 days.

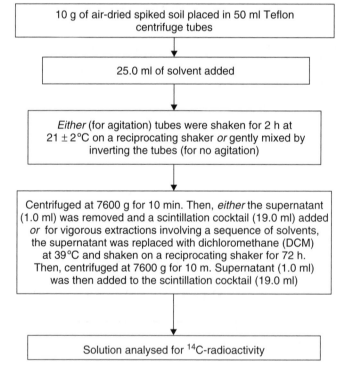

Figure 5.2 Procedure adopted in the 'mild-solvent' extraction method used by Kelsey *et al.* [3].

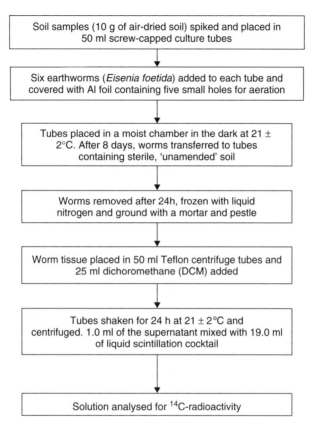

Figure 5.3 Procedure adopted in the earthworm uptake studies used by Kelsey *et al.* [3].

DQ 5.4

What do you observe about the results shown in Table 5.3?

Answer

It is noted that all of the solvents recovered smaller quantities with respect to the aging process. Sequential (vigorous) extraction with *n*-butanol, followed by dichloromethane, produced quantitative recovery at 120 days, as well as from the freshly spiked soil. Results determined by HPLC indicated the presence of phenanthrene as the radioactive component throughout the experiment. It was noted that the uptake by earthworms most closely resembled the data obtained when using *n*-butanol 'without agitation', whereas bacterial mineralization most closely resembled the *n*-butanol data, 'with agitation'. Overall, the data indicate

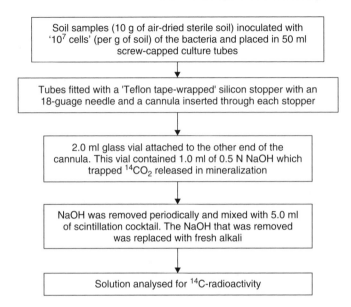

Figure 5.4 Procedure adopted in the bacterial mineralization studies, employing the *Pseudomonas* strain R for ^{14}C-labelled phenanthrene, used by Hatzinger and Alexander [2] and Kelsey *et al.* [3].

Table 5.3 Influence of solvent type on the recovery of phenanthrene from aged soils: comparison with earthworm uptake and bacterial mineralization studios [3]

Extraction solvent	Bioavailable or extractable content at 'day-0' (%)[a]	Bioavailable or extractable content relative to 'day-0' value (%)		
		7 days[b]	50 days[b]	120 days
n-Butanol	83.3	85.7	67.5	61.5
n-Butanol (no agitation)	62.1	62.9	45.2	19.5
Acetonitrile–water (1:1)	64.2	90.5	67.3	71.0
Ethanol–water (35:65)	7.99	78.1	73.3	58.9
Ethanol–water (1:1)	12.6	80.9	72.9	70.5
Ethanol–water (3:2)	29.5	85.5	72.8	66.4
Methanol–water (45:55) at 40°C	12.2	86.8	84.4	57.9
Methanol–water (45:55)	5.6	75.7	78.7	60.7
Methanol–water (1:1)	11.5	77.0	70.2	63.7
n-Butanol followed by dichloromethane	108	ND	ND	94.4
Bioassay				
Earthworm uptake	8.7	65.7	56.5	29.6
Mineralization	23.0	80.4	64.3	60.0

[a] Amount available to test organism or that which was extracted.
[b] ND, not determined.

that a procedure to predict bioavailability based on selective, 'mild-solvent' extraction using butan-1-ol is feasible.

DQ 5.5

What is the disadvantage of liquid scintillation counting?

Answer
The disadvantage of this approach is that it is the radioactive label which is being identified and not the specific-labelled compound.

5.2.1.1 An Approach Based on the Hildebrand Solubility Parameter
An advanced approach to select organic solvents that may act as 'strong' solvents and 'weak' solvents has been evaluated. This approach is based on the Hildebrand (total) solubility parameter (δ_t) [6]. The Hildebrand solubility parameter is a measure of the internal energy of cohesion in the solvent/analyte and can be represented as follows:

$$\delta_t = (\Delta E_v / V)^{1/2} \tag{5.1}$$

where δ_t is the *total* Hildebrand solubility parameter, ΔE_v is the energy of vaporization at a given temperature and V is the molar volume of the molecule.

This approach can be expanded to include contributions from interactions commonly found in matter, i.e. polarity, dispersion and hydrogen-bonding ability. These properties can be linked to the total Hildebrand solubility parameter as follows:

$$\delta_t^2 = \delta_p^2 + \delta_d^2 + \delta_h^2 \tag{5.2}$$

After calculating the values of the individual parameters for POPs and solvents, these can be plotted. Use of this approach allows solvents to be selected which are either effective (for total POP recovery) or selective (for determining the potential 'bioavailable' fraction) from soil matrices.

Calculation of Solvent/POP Parameters Thermodynamic data are available in the scientific literature to enable calculation of the individual parameters [7] of the total solubility parameter (Table 5.4). The approach [8] involves selecting data to determine each group's contribution to polarity, dispersion and hydrogen bonding (i.e. F_p, F_d and U_h, respectively). Then, by considering the structure of each solvent/POP under investigation, it is possible to determine the group contributions. This is demonstrated for a solvent, i.e. methanol, and a POP, i.e. 4,4'-DDT, based on their chemical structures (Figure 5.5). In the case of methanol, this is carried out as follows. The chemical formula of methanol consists of one methyl group and one hydroxyl group. By using the data in Table 5.4, it is possible

Table 5.4 Selected fractional parameters for dispersion, polarity and hydrogen bonding [7]

Molecular fragment	Fractional parameter for dispersion, F_d (kJ)	Fractional parameter for polarity, F_p (kJ)	Fractional parameter for hydrogen bonding, F_h (J)	Molar volume, V (cm^3 mol^{-1})
–CH$_3$	420	0	0	33.5
–CH$_2$	270	0	0	16.1
>CH–	80	0	0	−1
>C<	−70	0	0	−19.2
=CH$_2$	400	0	0	28.5
=CH–	200	0	0	13.5
=C<	70	0	0	−5.5
Ph	0	0	0	71.4
Ph–	1430	110	0	52.4
–Ph–	1270	110	0	33.4
–F	220	0	0	18
–Cl	450	550	400	24
–Br	550	0	0	30
–CN	430	1100	2500	24
–OH	210	500	20 000	10
–O–	100	400	3000	3.8
–CO–	290	770	2000	10.8
–COO–	390	490	7000	18
–NH$_2$	280	0	8400	19.2
–NH–	160	210	3100	4.5
–N<	20	800	5000	−9
1 plane of symmetry	—	0.5×	—	—
2 planes of symmetry	—	0.25×	—	—
More planes of symmetry	—	0×	—	—

to identify values for the CH$_3$ and OH groups and their contributions to dispersion (F_d), polarity (F_p), hydrogen bonding (U_h) and molar volume (V). By summing the values, it is possible to determine the 'individual group contributions' (F_d, F_p and U_h) and the molar volume (V) (see Table 5.5). The same approach can be applied to a POP, in this case 4,4′-DDT. By using the structure for this POP shown in Figure 5.5, it is possible to identify the key chemical fragments (Table 5.4) to calculate the 'individual group contributions' (Table 5.5). Once the latter contributions have been calculated, the values are inserted into the following equations (5.3–5.6) to determine the hydrogen-bonding ability, δ_h, the dispersion-coefficient contribution, δ_d, and the polarity contribution, δ_p

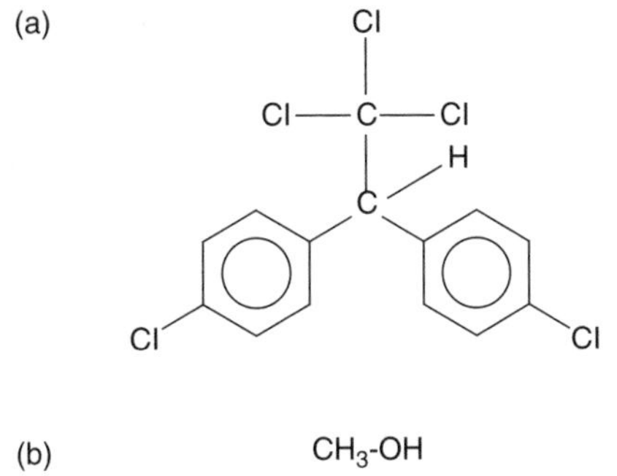

(a)

(b) CH$_3$-OH

Figure 5.5 Molecular structures of (a) 1,1,1-trichloro-2,2-bis(p-chlorophenyl)ethane (4,4′-DDT), and (b) methanol.

Table 5.5 Individual group contributions for methanol and 4,4′-DDT

Compound	Group	Group contribution to dispersion, F_d ($J^{1/2}$ cm$^{3/2}$ mol^{-1})	Group contribution to polarity, F_p ($J^{1/2}$ cm^2 mol^{-1})	Group contribution to hydrogen bonding, U_h (J mol^{-1})	Molar volume, V (cm^3 mol^{-1})
Methanol	CH$_3$	420	0	0	33.5
	OH	210	500	20 000	10.0
	Total	*630*	*500*	*20 000*	*43.5*
4, 4′ − *DDT*	2 × –Ph–	2540	48 400	0	104.8
	2 × Cl–CH=	900	1 210 000	800	48
	3 × Cl	1350	2 722 500	1200	72
	1 × CH	80	0	0	−1.0
	>C<	−70	0	0	−19.2
	Total	*4800*	*3 980 900*	*2000*	*204.6*

(Figures 5.6 and 5.7 for methanol and 4,4′-DDT, respectively):

$$\delta_d = (\Sigma F_d)/V \tag{5.3}$$

$$\delta_p = (\Sigma F_p)/V^{\ddagger} \tag{5.4}$$

$$\delta_p = (\Sigma F_p^2)^{1/2}/V \tag{5.5}$$

$$\delta_h = ((\Sigma U_h)/V)^{1/2} \tag{5.6}$$

The total Hildebrand solubility parameter, δ_t, can be calculated by using equation (5.2). A summary of the calculated values for hydrogen-bonding ability,

---

$\ddagger$ For molecules with more than one polar group present, then equation (5.5) must be used instead of equation (5.4), to take into account the interactions between the polar groups.

(a) *Calculation of the dispersion coefficient, δ_d, for methanol*

- Equation (5.3) $\delta_d = \dfrac{(\sum F_d)}{V}$

- Total F_d (Table 5.5) $630\,J^{1/2}\,cm^{3/2}\,mol^{-1}$

- Divide by V (Table 5.5) $\dfrac{630}{43.5}\ \dfrac{J^{1/2}\,cm^{3/2}}{cm^3}$

  $= 14.48\,J^{1/2}\,cm^{-3/2}$
  $= 14.48\,(J\,cm^{-3})^{1/2}$
  $= 14.48\,(MPa)^{1/2}$

(b) *Calculation of the polarity contribution, δ_p, for methanol*

- Equation (5.4) $\delta_p = \dfrac{(\sum F_p)}{V}$

- Total δ_p (Table 5.5) $500\,J^{1/2}\,cm^2\,mol^{-1}$

- Divide by V (Table 5.5) $\dfrac{500}{43.5}\ \dfrac{J^{1/2}\,cm^2}{cm^3}$

  $= 11.49\,J^{1/2}\,cm^{-3/2}$
  $= 11.49\,(J\,cm^{-3})^{1/2}$
  $= 11.49\,(MPa)^{1/2}$

(c) *Calculation of the hydrogen-bonding contribution, δ_h, for methanol*

- Equation (5.6) $\delta_h = \dfrac{(\sum U_h)^{1/2}}{V}$

- Total δ_h (Table 5.5) $20\,000\,J^{1/2}\,mol^{-1}$

- Divide by V (Table 5.5) $\dfrac{20\,000}{43.5}\ \dfrac{J^{1/2}}{cm^3}$

- Take square root $= 459.77\,J^{1/2}\,cm^{-3/2}$
  $= 21.44\,(J\,cm^{-3})^{1/2}$
  $= 21.44\,(MPa)^{1/2}$

Figure 5.6 An approach adopted for the calculation of the dispersion coefficient, polarity contribution and hydrogen-bonding ability of methanol.

δ_h, the dispersion-coefficient contribution, δ_d, and the polarity contribution, δ_p, together with the total Hildebrand solubility parameter, δ_t, for a range of solvents and POPs is shown in Table 5.6.

SAQ 5.3

Calculate the total Hildebrand solubility parameter, δ_t for the solvents acetonitrile and dichloromethane by using equation (5.2) and the data shown in Table 5.6.

(a) *Calculation of the dispersion coefficient, δ_d, for 4,4'-DDT*

- Equation (5.3)

$$\delta_d = \frac{(\sum F_d)}{V}$$

- Total F_d (Table 5.5)

$4800 \, J^{1/2} \, cm^{3/2} \, mol^{-1}$

- Divide by V (Table 5.5)

$$\frac{4800 \quad J^{1/2} \, cm^{3/2}}{204.6 \quad cm^3}$$

$= 23.46 \, J^{1/2} \, cm^{-3/2}$
$= 23.46 \, (J \, cm^{-3})^{1/2}$
$= 23.46 \, (MPa)^{1/2}$

(b) *Calculation of the polarity contribution, δ_p, for 4,4'-DDT*

- Equation (5.5)

$$\delta_p = \frac{(\sum F_p)^{1/2}}{V}$$

- Square each group contribution

$48\,400 \, J^{1/2} \, cm^{3/2} \, mol^{-1}$
$1\,210\,000 \, J^{1/2} \, cm^{3/2} \, mol^{-1}$
$2\,722\,500 \, J^{1/2} \, cm^{3/2} \, mol^{-1}$

- Add up squared contributions
- Take square root

$3\,980\,900 \, J \, cm^3 \, mol^{-2}$
$1995 \, J^{1/2} \, cm^{3/2} \, mol^{-1}$

- Divide by V (Table 5.5)

$$\frac{1995 \quad J^{1/2} cm^{3/2}}{204.6 \quad cm^3}$$

$= 9.75 \, J^{1/2} \, cm^{-3/2}$
$= 9.75 \, (J \, cm^{-3})^{1/2}$
$= 9.75 \, (MPa)^{1/2}$

(c) *Calculation of the hydrogen-bonding contribution, δ_h, for 4,4'-DDT*

- Equation (5.6)

$$\delta_h = \frac{(\sum U_h)^{1/2}}{V}$$

- Total δ_h (Table 5.5)

$2000 \, J^{1/2} \, mol^{-1}$

- Divide by V (Table 5.5)

$$\frac{2000 \quad J^{1/2}}{204.6 \quad cm^3}$$

$= 9.75 \, J^{1/2} \, cm^{-3/2}$

- Take square root

$= 3.13 \, (J \, cm^{-3})^{1/2}$
$= 3.13 \, (MPa)^{1/2}$

Figure 5.7 An approach adopted for the calculation of the dispersion coefficient, polarity contribution and hydrogen-bonding ability of 4,4'-DDT.

SAQ 5.4

Calculate the total Hildebrand solubility parameter, δ_t for the persistent organic pollutants, lindane and hexachlorobenzene, by using equation (5.2) and the data shown in Table 5.6.

Table 5.6 Summary of selected total hildebrand solubility parameters and their individual components for solvents and POPs

Material	Dispersion coefficient, δ_d (MPa$^{1/2}$)	Polarity, δ_p (MPa$^{1/2}$)	Hydrogen bonding, δ_h (MPa$^{1/2}$)	Total Hildebrand solubility parameter, δ_t (MPa$^{1/2}$)
Solvents				
Methanol	14.5	11.5	21.4	28.3
Toluene	17.6	1.1	0	17.6
Acetonitrile	14.8	19.1	6.6	See SAQ 5.3
Acetone	14.5	9.9	5.1	18.3
Dichloromethane	18.3	8.6	3.5	See SAQ 5.3
Isohexane	14.3	0	0	14.3
Heptane	14.9	0	0	14.9
Acetonitrile:dichloromethane (1:1) (vol/vol)	16.5	13.9	5.1	22.2
POPs				
4,4′-DDT	23.5	9.8	3.1	25.6
4,4′-DDD	21.6	7.8	2.8	23.1
4,4′-DDE	25.0	8.3	2.9	26.5
PCP	35.1	26.3	14.4	46.2
Lindane	21.0	16.8	3.5	See SAQ 5.4
Endosulfan	24.5	17.6	7.5	31.1
Endrin	24.4	14.3	4.8	28.7
2-Chlorophenol	28.6	13.8	17.4	33.5
Naphthalene	27.3	2.1	0	27.4
Nitrobenzene	22.9	12.7	4.2	23.3
Hexachlorobenzene	23.3	18.6	3.7	See SAQ 5.4
2,4-Dichlorophenol	26.0	13.3	15.1	32.9
Fluorene	33.9	2.4	0	34.0
Fluoranthene	38.0	3.0	0	38.1

Note: For solvent mixtures, the following approach should be used to calculate the individual parameters. For example, the polarity contribution, δ_p, for a 1:1 (vol/vol) ratio of acetonitrile and dichloromethane is calculated as follows: $\delta_p = (0.5\delta_p(\text{acetonitrile}) + 0.5\delta_p(\text{dichloromethane}))$, whereas for a 1:2 (vol/vol) ratio of acetonitrile and dichloromethane it is calculated as follows: $\delta_p = (0.33\delta_p(\text{acetonitrile}) + 0.67\delta_p(\text{dichloromethane}))$. Similarly, the dispersion coefficient contribution, δ_d, and the hydrogen-bonding ability, δ_h, can be calculated.

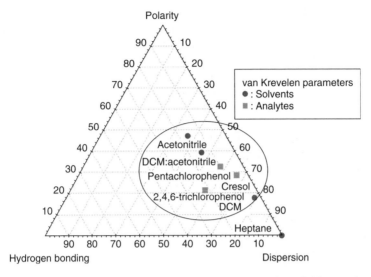

Figure 5.8 Solvent predictions for phenols in CRM401-225 (DCM, dichloromethane) [9].

SAQ 5.5

Calculate the dispersion coefficient for a 1:4 (vol/vol) ratio of acetonitrile and dichloromethane.

Visualization of the Calculated Parameters Visualization of the hydrogen bonding ability, δ_h, the dispersion-coefficient contribution, δ_d, and the polarity contribution, δ_p, can be carried out by normalizing them with respect to the total Hildebrand solubility parameter, using equations (5.7–5.9), and then plotting them on a triangular graph (see, for example, Figure 5.8):

$$(\delta_d^2/\delta_t^2) \times 100 = Fr_d \tag{5.7}$$

$$(\delta_p^2/\delta_t^2) \times 100 = Fr_p \tag{5.8}$$

$$(\delta_h^2/\delta_t^2) \times 100 = Fr_h \tag{5.9}$$

Case Study 1: Sequential Extraction of POPs from a Certified Reference Material, CRM401-225[†]

This approach to assess the sequential extraction of POPs from Certified Reference Materials was carried out by using pressurized fluid extraction. Extraction

---

[†] See Scott and Dean [9].

conditions were fixed at a temperature of 100°C and a pressure of 2000 psi (1 psi = 6894.76 Pa), with an extraction time of 10 min (i.e. 5 min for temperature equilibration and 5 min for static extraction). Only one static flush cycle was used. The extraction cell was then flushed with fresh solvent and N_2. The total extraction time was approximately 13 min per sample. Each extract was then transferred to a 25.0 ml volumetric flask, an internal standard (50 μl of pentachloronitrobenzene) was added and each flask 'filled to the mark' with solvent.

Analysis was carried out by using a GC–MS system (HP G1800A GCD, Hewlett Packard, Palo Alto, CA, USA) operated in the single-ion monitoring mode with a splitless injection volume of 1 μl. The column was a DB-5ms (purchased from Sigma-Aldrich Company Ltd, Gillingham, Dorset, UK) with the following dimensions: length 30 m × 0.25 mm internal diameter × 0.25 μm film thickness. For all analyses, the injection port temperature was set at 250°C and the detector temperature was set at 280°C. Selected standards for the analytes were run daily to assess analytical performance. The temperature programme used for separation of the phenols was as follows: initial temperature 90°C for 2 min and then to 230°C at 7°C min^{-1}. Separation of all of the target phenols was achieved in approximately 22 min.

In this case study, solvent selection was applied for the sequential extraction of three POPs, i.e. pentachlorophenol, total cresol and 2,4,6-trichlorophenol, from a soil Certified Reference Material (CRM401-225).

DQ 5.6

Using the Hildebrand solubility parameter approach (Figure 5.8), which types of solvents would you class heptane and acetonitrile as for recovering pentachlorophenol, total cresol and 2,4,6-trichlorophenol from soil?

Answer

Using this approach, heptane would be classed as a weak solvent for extracting these POPs whereas acetonitrile would be a strong solvent.

In order to demonstrate this approach, separate sub-samples of the CRM were extracted by using pressurized fluid extraction, as described above, using heptane. It was found that the predicted weak solvent, i.e. heptane, was able to recover approximately 70% of the phenols (Figure 5.9). The same sample was left in the extraction cell and re-extracted using a stronger solvent, i.e. acetonitrile. The latter was able to recover an additional approximate 30% of phenols from the soil.

DQ 5.7

Comment on the results shown in Figure 5.9, considering, in particular, whether results are in agreement with the certificate values and total

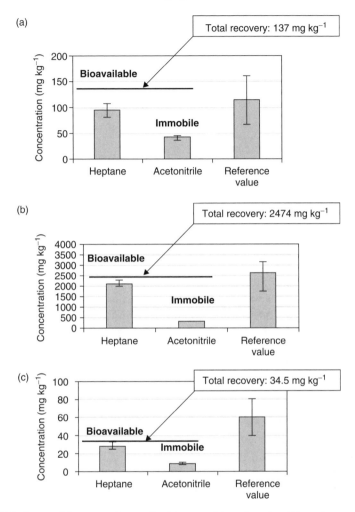

Figure 5.9 Sequential extraction of (a) pentachlorophenol, (b) total cresol and (c) 2,4,6-trichlorophenol from CRM401-225. Recoveries for heptane and acetonitrile are quoted on the basis of the mean ± standard deviation (SD) for $n = 5$. The certificate (reference) values are also quoted on the basis of the mean ± standard deviation (SD) [9].

recoveries using weak and strong solvents for pentachlorophenol, total cresol and 2,4,6-trichlorophenol from soil.

Answer

In the case of pentachlorophenol (Figure 5.9(a)), the total recovery by both solvents (137 mg kg^{-1}) was in agreement with the certificate value but slightly lower than that obtained by acetonitrile only (Table 5.7).

Table 5.7 Recovery of POPs from Certified Reference Materials [9]

Recovery process	Certified Reference Material	Compound	Certificate value ± SD (mg kg^{-1})	Mean recovery ± SD (mg kg^{-1})^a
Extraction of phenols using acetonitrile	CRM401-225	Cresol (total)	2657.8 ± 889.3	1815.6 ± 82.46
		2,4,6-Trichlorophenol	58.7 ± 19.4	40.93 ± 5.12
		Pentachlorophenol	117.1 ± 41.7	150.6 ± 8.44
Extraction of pesticides using acetone	CRM805-050	Lindane	10.6 ± 4.85	9.34 ± 0.64
		Endosulfan I	6.9 ± 3.79	5.97 ± 0.6
		p,p'- DDE	18.6 ± 9.10	12.02 ± 1.41
		Endrin	12.97 ± 7.51	13.33 ± 0.46
		p,p'-DDD	19.45 ± 8.64	20.40 ± 1.42

^{a} For $n = 5$.

For cresol (total) (Figure 5.9(b)), the total recovery by both solvents (2474 mg kg^{-1}) was in agreement with the certificate value but exceeded the recovery obtained by acetonitrile only (Table 5.7). For 2,4,6-trichlorophenol (Figure 5.9(c)), the total recovery by both solvents (34.5 mg kg^{-1}) was lower than the certificate value but in agreement with the recovery obtained by acetonitrile only (Table 5.7).

Case Study 2: Sequential Extraction of POPs from a Certified Reference Material, CRM805-050[‡]

This approach to assess the sequential extraction of POPs from certified reference materials was carried out by using pressurized fluid extraction employing the same approach as described in Case Study 1. Analysis was achieved by using GC-MS (details described in Case Study 1), with the following temperature programme: initial temperature 120°C for 2 min and then to 250°C at 4°C min^{-1}. Separation of all of the target pesticides was achieved in approximately 34 min.

In this case study, a range of pesticides were sequentially extracted from a soil Certified Reference Material (CRM805-050).

SAQ 5.6

Using the Hildebrand solubility parameter approach (Figure 5.10), which types of solvents would you class toluene and acetone as for recovering lindane, endosulfan I, *p,p'*-DDE, endrin and *p,p'*-DDD from soil?

[‡] See Scott and Dean [9].

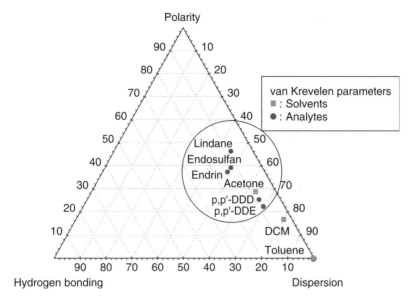

Figure 5.10 Solvent predictions for CRM805-050 (DCM, dichloromethane) [9].

The results, shown in Figure 5.11, indicate that exhaustive extraction with a weak solvent liberated approximately 70% of the POPs, with total recovery being achieved when the sample was then extracted with the strong solvent.

SAQ 5.7

Comment on the results shown in Figure 5.11, considering, in particular, whether results are in agreement with the certificate values and total recoveries using weak and strong solvents for lindane, endosulfan I, p,p'-DDE, endrin and p,p'-DDD from soil.

Using these case studies, it has been shown that it is possible (a) to predict weak and strong solvents using the Hildebrand solubility parameter approach, and (b) fractionate POPs using selective solvent extraction. Labelling of the fractions as 'bioavailable fraction' and 'immobile fraction' is unproven and further work is required to investigate the relationship between the fractionation of POPs from soil and their uptake by micro-organisms, plants and animals. In addition, the approach described does not take into account the interactions that the POPs will have with the soil matrix and the possible degradation products that may arise due to microbial breakdown on POPs in the soil environment.

5.2.2 Cyclodextrin Extraction

Cyclodextrins have been used to predict PAH bioavailability in contaminated samples.

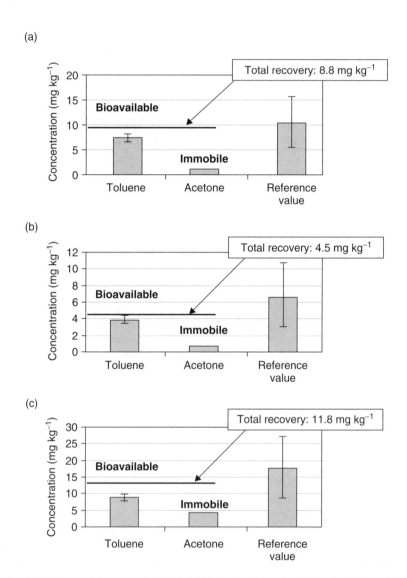

Figure 5.11 Sequential extraction of (a) lindane, (b) endosulfan 1, (c) p, p'-DDE, (d) endrin, and (f) p, p'-DDD from CRM805-050. Recoveries for toluene and acetone are quoted on the basis of the mean $\pm$ standard deviation (SD) for $n = 5$. The certificate (reference) values are also quoted on the basis of the mean $\pm$ standard deviation (SD) [9].

(d)

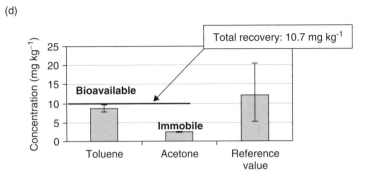

(e)

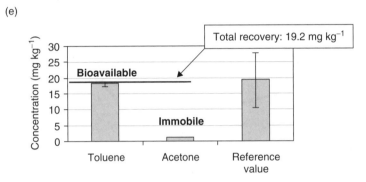

Figure 5.11 (*continued*).

DQ 5.8

What are cyclodextrins?

Answer

Essentially, cyclodextrins are cyclic oligiosaccharides with a hydrophilic shell and a toroidal-shaped, apolar (hydrophobic) cavity (Figure 5.12).

Their importance in bioavailability studies is that they can form water-soluble inclusion complexes with hydrophobic organic molecules, i.e. mobile polycyclic (polynuclear) aromatic hydrocarbons (PAHs) present in soil can be encapsulated by the cyclodextrin macromolecule, thus forming an inclusion complex. It is therefore a good assumption that organic compounds which are strongly bound or sequestered in the soil are not in the soil-pore water and hence cannot be extracted by using cyclodextrin. This approach is valid, provided that the shape and size of the organic compounds under investigation are compatible with the cyclodextrin cavity. Table 5.8 indicates the size of the cyclodextrin cavity for

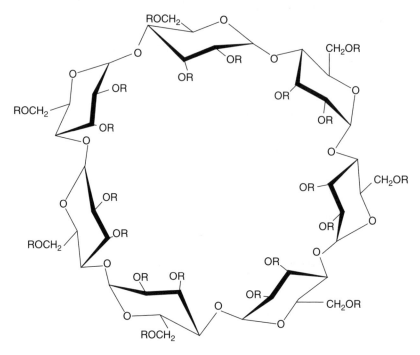

Figure 5.12 Molecular structures of β-cyclodextrin (R=H) and hydroxyl β-cyclodextrin (HPCD) (R=CH₂CH(OH)CH₃).

Table 5.8 Cyclodextrin and PAH compatibility data

Compound	Size (width–length)(nm)	Reference
α-Cyclodextrin	0.57–0.78	Szejtli [10]
β-Cyclodextrin	0.78–0.78	Szejtli [10]
γ-Cyclodextrin	0.95–0.78	Szejtli [10]
Naphthalene	0.5–0.71	Wang and Brusseau [11]
Phenanthrene	0.58–0.78	Wang and Brusseau [11]
Anthracene	0.5–0.92	Wang and Brusseau [11]
Fluoranthene	0.71–0.92	Wang and Brusseau [11]
Pyrene	0.71–0.89	Wang and Brusseau [11]

α-cyclodextrin (which comprises six D-glucose units) and β-cyclodextrin (which comprises seven D-glucose units).

DQ 5.9

How many D-glucose units would you expect to find in γ-cyclodextrin?

Answer

γ-Cyclodextrin comprises eight D-glucose units.

Table 5.8 also shows the sizes of various PAHs. Based on this information, it is reasonable to assume that naphthalene and anthracene will form 1:1 inclusion complexes with β-cyclodextrin, while pyrene, acenaphthene, phenanthrene and fluoranthene form 1:2 complexes with β-cyclodextrin [11]. In reality, the most popular form of cyclodextrin that has been used is hydroxypropyl-β-cyclodextrin (HPCD) (Figure 5.12). A procedure for the extraction of ^{14}C-radiochemically labelled PAHs is shown in Figure 5.13 [12, 13].

Recent data have looked at the dependency of soil properties on the recovery of ^{14}C-labelled phenanthrene[§] from aged soils with different properties [13]. The soil properties are shown in Table 5.9, indicating differences in clay content (soils A and B versus soil C), sand content (soil A versus soils B and C), silt content (all three soils), loss on ignition (soils A and B versus soil C) and pH (soils A and B versus soil C). Data for the recovery of ^{14}C-labelled phenanthrene from the soil are shown in Table 5.10. It is readily seen, at 'day 1', that the recovery of phenanthrene is unaffected by soils A and B (low clay content), whereas approximately 50% loss of phenanthrene had occurred from soil C (20% clay

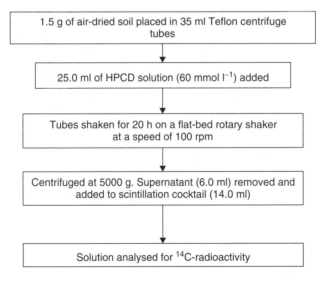

Figure 5.13 Procedure adopted in the extraction of ^{14}C-radiochemically labelled PAHs using hydroxypropyl-β-cyclodextrin (HPCD) [12, 13].

[§] The use of ^{14}C-labelled compounds, and subsequent measurement by liquid scintillation counting, allows an activity mass balance to be carried out, ensuring that all of the labelled compound can be accounted for. The disadvantage of this approach is that it is the radioactive label which is being identified and not the specifically labelled compound.

Table 5.9 Soil properties for phenanthrene recovery using HPCD [13]

Soil identifier	Clay (%)	Sand (%)	Silt (%)	pH	Loss on ignition (%)	Texture
Soil A	2	88	10	7.1	2.4	Sandy loam
Soil B	9	47	44	7.2	3.1	Sandy silt loam
Soil C	20	53	27	4.7	39.7	Loamy peat

Table 5.10 Recovery of ^{14}C-phenanthrene from aged soils [13]

Soil aging	Soil A[a]		Soil B[a]		Soil C[a]	
	HPCD recovery (%)	BuOH recovery (%)[b]	HPCD recovery (%)	BuOH recovery (%)[b]	HPCD recovery (%)	BuOH recovery (%)[b]
1 day	97.9 ± 8.1	90.4 ± 8.2	100 ± 9.2	84.9 ± 7.5	53.8 ± 17.8	69.3 ± 16.6
40 days	11.0 ± 2.2	9.6 ± 2.3	10.4 ± 4.2	17.6 ± 7.3	40.3 ± 5.3	47.6 ± 5.0
80 days	12.5 ± 14.0	6.5 ± 1.1	9.5 ± 4.1	18.6 ± 4.6	10.9 ± 6.1	10.7 ± 1.4

[a] Mean values $\pm 95\%$ confidence intervals.
[b] Data reported at t_{rap}, i.e. extraction time required to extract the rapidly desorbing fraction. Experimental values varied between 49.7 and 50.5 s for soil A, 19.8 and 52.2 s for soil B and 22.9 and 48.4 s for soil C.

content). After 40 days of aging, phenanthrene is poorly recovered from soils A and B (ca. 10%), while for soil C ca. 40% is recoverable. Similar findings are reported for soils A and B at 80 days of aging, while a corresponding loss of recovery is also noted for soil C (ca. 10% recovered). The data were compared to that obtained when using butan-1-ol (see Section 5.2.1). It was concluded that while some correlation exists between the two approaches, further work was required to assess compound bioavailability.

5.2.3 Supercritical-Fluid Extraction

Supercritical-fluid extraction (SFE) relies on the diversity of properties exhibited by the supercritical fluid to (selectively) extract analytes from solid, 'semi-solid' or liquid matrices.

DQ 5.10

What are the important properties of a supercritical fluid?

Answer

The important properties offered by a supercritical fluid for extraction are (a) good solvating power, (b) high diffusivity and low viscosity, and (c) minimal surface tension.

Figure 5.14 shows a phase diagram for a pure substance which identifies the regions where the substance occurs, as a consequence of temperature or pressure, as a single phase, i.e. solid, liquid or gas. The divisions between these regions are bounded by curves indicating the co-existence of three phases, namely solid–gas corresponding to sublimation, solid–liquid corresponding to melting and liquid–gas corresponding to vaporization. The three curves intersect at the triple point where the three phases co-exist in equilibrium. At the critical point, designated by both a critical temperature and a critical pressure, no liquefaction will take place on raising the pressure and no gas will be formed on increasing the temperature. It is this specific region, which is, by definition, the *supercritical* region. Table 5.11 shows the critical temperatures and pressures of a range of substances which are suitable for SFE. The most common supercritical fluid is CO_2.

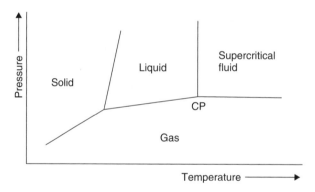

Figure 5.14 Phase diagram for a pure substance (CP, critical point) [41]. From Dean, J. R., *Extraction Methods for Environmental Analysis*, Copyright 1998. © John Wiley & Sons, Limited. Reproduced with permission.

Table 5.11 Critical properties of some selected substances [14]

Substance	Critical temperature (°C)	Critical pressure	
		atm	psi
Ammonia	132.4	115.0	1646.2
Carbon dioxide	31.1	74.8	1070.4
Chlorodifluoromethane	96.3	50.3	720.8
Ethane	32.4	49.5	707.8
Methanol	240.1	82.0	1173.4
Nitrous oxide	36.6	73.4	1050.1
Water	374.4	224.1	3208.2
Xenon	16.7	59.2	847.0

DQ 5.11

Does carbon dioxide have a permanent dipole moment?

Answer

Carbon dioxide has no permanent dipole moment.

As a consequence of this lack of a permanent dipole moment, CO_2 has no polarity. However, by addition of a polar organic solvent (or modifier), it is possible to increase the flexibility of the system.

The basic components of an SFE system (Figure 5.15) are a supply of high-purity carbon dioxide, a supply of high-purity organic modifier, two pumps, an oven for the extraction vessel, a pressure outlet or restrictor and a suitable collection vessel for quantitative recovery of extracted analytes.

SFE has been applied as a selective technique to investigate the desorption of organic pollutants including polychlorinated biphenyls (PCBs) [15, 16] and poly-cyclic (polynuclear) aromatic hydrocarbons (PAHs) [17] from a range of environmental matrices. Initial work using supercritical CO_2 from 40 to $150°C$ [15] developed a sequential extraction process for PCBs that was able to define four stages of the extraction process. These stages were (1) rapidly desorbing, (2) moderately desorbing, (3) slowly desorbing and (4) very slowly desorbing, corresponding to extraction conditions of 120 bar at $40°C$, 400 bar at $40°C$, 400

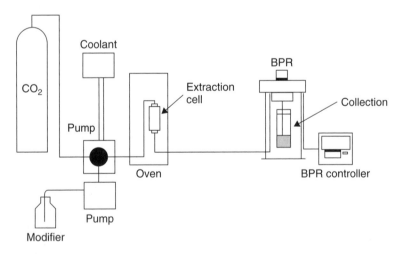

Figure 5.15 Schematic diagram of the basic components of a supercritical fluid extraction system (BPR, back-pressure regulator (restrictor)) [41]. From Dean, J. R., *Extraction Methods for Environmental Analysis*, Copyright 1998. © John Wiley & Sons, Limited. Reproduced with permission.

bar at 100°C and 400 bar at 150°C, respectively (extraction for 1 h at each stage). Initially, however, the individual PCB congeners were extracted from Certified Reference Materials for total recovery using SFE (400 bar at 150°C for 1 h). Table 5.12 demonstrates the compliance with certified values for the recovery of PCB congeners. The distribution of PCB congeners following selective SFE is shown in Table 5.13.

Table 5.12 Recovery of PCB congeners from environmental Certified Reference Materials using supercritical CO_2 (400 bar at 150°C for 1 h) [15]

CRM[a]	PCB congener	Concentration determined (ng g^{-1})[b]	Certificate value (ng g^{-1})
CRM536	28	46.8 (2.6)	44.4 ± 6.3
	52	34.5 (0.9)	38.4 ± 7.1
	101	37.9 (2.4)	43.7 ± 6.2[c]
	118	25.7 (2.6)	27.6 ± 4.0
	138	31.6 (2.5)	26.8 ± 3.9
	149	43.4 (1.9)	48.8 ± 5.9
	153	45.7 (1.9)	50.3 ± 5.7
	170	10.9 (1.5)	13.4 ± 1.5
	180	18.9 (5.3)	22.4 ± 3.6
SRM1944	28	69.5 (5.5)	75.8 ± 2.2
	52	52.2 (6.0)	78.9 ± 1.7
	101	44.0 (4.5)	73.3 ± 1.5
	105	18.7 (2.8)	22.4 ± 0.8
	118	41.8 (3.7)	57.6 ± 1.3
	138	38.8 (3.2)	59.7 ± 1.5
	149	36.2 (4.4)	49.1 ± 1.7
	153	44.2 (4.9)	73.5 ± 1.5
CRM481	52	1670 (2.3)	2900 ± 440[c]
	101	27 200 (2.9)	37 000 ± 4800
	118	7310 (1.7)	9400 ± 1100
	138	71 100 (2.3)	92 000 ± 14000[c]
	149	72 200 (1.7)	97 000 ± 9700
	153	107 000 (2.6)	137 000 ± 12000
	180	99 100 (1.5)	124 000 ± 11000
SRM1939	52	2970 (3.3)	4320 ± 130
	101	448 (2.5)	NA[d]
	138	216 (1.8)	258 ± 7
	149	349 (2.9)	427 ± 17
	153	228 (3.5)	297 ± 19

[a] CRM536, harbour sediment; SRM1944, marine sediment; CRM481, industrial soil; SRM1939, river sediment.
[b] Relative standard deviation (RSD) values (%) shown in parentheses.
[c] Indicative value only.
[d] NA, not available.

Table 5.13 Distribution of PCB congeners[a] after selective SFE from environmental Certified Reference Materials[b] [15]

Time (min)	0–60	60–120	120–180	180–240
Fraction	Fast	Moderate	Slow	Very slow
SFE	120 bar at	400 bar at	400 bar at	400 bar at
conditions	40°C	40°C	100°C	150°C
CRM536	46–70	7–26	15–19	5–10
SRM1944	72–89	3–10	4–16	3–7
CRM481	68–81	5–18	8–11	5–7
SRM1939	43–48	10–14	24–30	13–17

[a] Values expressed as percentage content (range) in each fraction.
[b] CRM536, harbour sediment; SRM1944, marine sediment; CRM481, industrial soil; SRM1939, river sediment.

DQ 5.12

Where are the highest recoveries obtained?

Answer

In all cases, the highest recoveries were obtained in stage (1), i.e. the PCB congeners were rapidly desorbed from their matrices.

The highest levels of very slowly desorbing (stage (4)) PCB congeners were found in CRM1939 (river sediment). This approach was then applied to estimate the fraction of PCB that is bioavailable to a benthic organism (chironomid larvae) in a naturally contaminated sediment [16]. Selective SFE at 120 bar and 40°C for 1 h was first applied to remove the PCBs from contaminated sediment. Then, chironomid larvae were added to separate samples of the contaminated sediment and pre-extracted sediment. The larvae were harvested after 2.5 months and extracted using SFE (355 bar at 100°C for 0.5 h). The results (Figure 5.16) for total PCB recovery indicate that the process of selective SFE on the original contaminated sediment ('treated') had removed a fraction of the available PCBs compared to the uptake by biota from the 'untreated' sediment. The authors [16] concluded that selective SFE (120 bar at 40°C for 1 h) was capable of removing bioavailable PCB congeners from contaminated sediment; furthermore, that selective SFE provides an alternative approach to assess bioavailability based on a chemical assay.

This approach (selective SFE) has also been compared to data obtained in field trials as part of a bioremediation study to assess the recovery of PAHs [17]. In this study, soil samples from a manufacturing gas plant were subjected to bioremediation over a 12 month period. Bioremediation involved 'feeding' the soil with nutrients and water as well as tilling frequently to enable aeration of the soil. Figure 5.17 indicates the amount of PAHs recovered after one year's

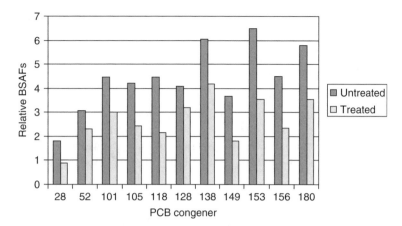

Figure 5.16 Relative biota-sediment accumulation factors (BSAFs) for polychlorinated biphenyls (PCBs) [16].

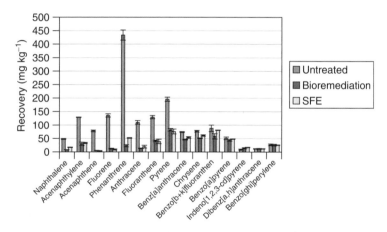

Figure 5.17 Comparison of polycyclic (polynuclear) aromatic hydrocarbon (PAH) recovery after bioremediation for one year with that estimated using selective supercritical-fluid extraction (performed under 'fast' (120 bar at 50°C for 1 h) conditions) [17]. Data displayed represent the mean values ± standard deviations (SDs) ($n = 4$ for bioremediation and SFE and $n = 3$ for untreated samples).

bioremediation compared to that recovered using selective SFE ('fast fraction' only, i.e. 120 bar at 50°C for 1 h). Concentration data values for the original untreated soil obtained after Soxhlet extraction for 18 h ($n = 3$) are included for comparison. It is noted (Figure 5.17) that a good correlation exists between the fast fraction obtained by selective SFE and the data reported for bioremediation after 1 year. The authors indicate that selective (or mild) SFE may provide a rapid and useful approach to assess the bioavailability of PAHs on contaminated soil.

5.2.4 Other Approaches

5.2.4.1 Sub-Critical Water Extraction (SWE)

A very similar approach to that described in Section 5.2.3 is to replace the CO_2 and use water as the extraction solvent. This approach involves varying the pressure and temperature of water to affect its polarity [18]. However, no specific data have been reported indicating its application to bioavailability studies.

5.2.4.2 Solid-Phase Microextraction (SPME)

In SPME, a fused-silica fibre coated with a stationary phase is exposed to the sample in its matrix (see also Chapter 3, Section 3.2.2). A typical stationary phase is polydimethylsiloxane although others are available (Table 5.14). The small size and cylindrical geometry allow the fibre to be incorporated into a syringe-type device (Figure 5.18). This allows the SPME device to be used for either GC or HPLC.

DQ 5.13

How can SPME be used for GC?

Answer

In GC, the SPME fibre simply replaces the syringe used to introduce samples into the hot injection port (where desorption of analytes takes place).

Table 5.14 Some commercially available SPME fibre coatings [14]. From Dean, J. R., *Methods for Environmental Trace Analysis*, ANTS Series. Copyright 2003. © John Wiley & Sons, Limited. Reproduced with permission

7 μm polydimethylsiloxane (bonded)
30 μm polydimethylsiloxane (non-bonded)
100 μm polydimethylsiloxane (non-bonded)
85 μm polyacrylate (partially cross-linked)
60 μm polydimethylsiloxane/divinylbenzene (partially cross-linked)
65 μm polydimethylsiloxane/divinylbenzene (partially cross-linked)
75 μm polydimethylsiloxane/Carboxen (partially cross-linked)
65 μm Carbowax/divinylbenzene (partially cross-linked)
50 μm Carbowax/Template resin (partially cross-linked)

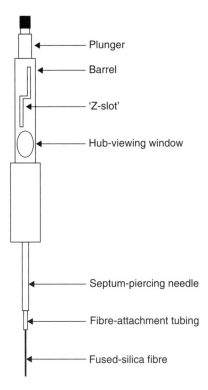

Figure 5.18 Schematic diagram of a solid-phase microextraction device. Reprinted with permission from Zhang, Z., Yang, M. and Pawliszyn, J., *Anal. Chem.*, **66**, 844A–853A (1994). Copyright (1994) American Chemical Society.

In HPLC, a modified injection manifold is required so that the mobile phase can be used to desorb the analytes. As can be seen in Figure 5.18, the fused-silica fibre (approximately 1 cm) is connected to a stainless-steel tube for mechanical strength. This assembly is mounted within the syringe barrel for protection when not in use. In operation, the SPME fibre is exposed to the sample (e.g. soil-pore water solution) for a pre-specified time. After a suitable time-duration, the fibre is withdrawn back into its protective syringe barrel and withdrawn from the sample and analysed. The potential of this approach to assess the 'unbound' fraction of POPs in soil pore waters is undeniable.

5.2.4.3 Membrane Separations

Semi-permeable membrane devices (SPMDs) are made from non-porous polyethylene tubing coated with a 'lipid' layer, e.g. triolein. This then allows contaminants in water to diffuse into the lipid coating according to the compounds' partition coefficients. This approach allows chemical exposure to be assessed directly.

DQ 5.14

What do you think is the major limitation of this type of device?

Answer

The major limitation of such devices is the dependence on partitioning alone to simulate a biological response.

5.3 Earthworm Studies

Another approach to assess the bioavailability of chemicals in soil is to use earthworms. For a full description of the types of earthworms and protocols, see Chapter 4, Section 4.4.

The use of earthworms to assess the bioavailability of POPs in soil requires the development of suitable extraction technology. Mooibroek *et al.* [19] compared a variety of extraction techniques for the determination of PAHs in earthworms. In their study, they compared a liquid-extraction technique based on light petroleum (LP) and microwave-assisted extraction (MAE) with a previously validated method [20] developed to extract total lipids in biological samples using non-chlorinated solvents. The three extraction methods are summarized in Table 5.15. Extracts were then 'cleaned-up' by using gel-permeation chromatography, followed by quantitation using reversed-phase liquid chromatography with fluorescence detection. The results are shown in Figure 5.19.

DQ 5.15

Comment on the standard deviations obtained for each PAH in Figure 5.19.

Answer

It is noted, in all cases, that the standard deviations for each PAH are high.

This is not surprising considering the type of biological sample used and the concentration level. The authors performed a statistical evaluation of the data and concluded that the 'light-petroleum method' gave similar results to that incurred by the standard method [20]. In the case of MAE, lower concentrations were determined for the less-volatile PAHs, namely dibenzo[ah]anthracene and benzo[ghi]perylene. In terms of lipid extraction, the 'light-petroleum', MAE and

Table 5.15 Comparison of extraction methods for the recovery of PAHs from earthworms [19][a]

Step	'Light-petroleum' method	MAE	'Smedes' method [20]
1	30 s with high-speed homogenizer with 20 ml light petroleum	10 min MAE with 10 ml isopropyl alcohol plus 10 ml cyclohexane	2 min with high-speed homogenizer with 9 ml isopropyl alcohol plus 10 ml cyclohexane
2	Extract over Na_2SO_4 into Kuderna–Danish apparatus[b]	Extract over Na_2SO_4 into Kuderna–Danish apparatus[b]	Addition of 10 ml water, followed by 1 min high-speed homogenization
3	30 s with high-speed homogenizer with 20 ml light petroleum	Evaporation of solvent	Centrifuge for 3 min
4	Evaporation of collected solvents to 1 ml volume	Re-dissolve in 2 ml of light petroleum	Pipette organic layer over Na_2SO_4 into Kuderna–Danish apparatus[b]
5	—	—	Addition of 10 ml isopropyl alcohol/cyclohexane, followed by 1 min high-speed homogenization
6	—	—	Centrifuge for 3 min
7	—	—	Pipette organic layer over Na_2SO_4 into Kuderna–Danish apparatus[b]
8	—	—	Evaporation of collected solvents
9	—	—	Re-dissolve in 2 ml of light petroleum

[a] All using 1 g of earthworm samples.
[b] See Dean [14].

'Smedes' [20] methods gave $0.86 \pm 0.12\%$, $1.93 \pm 0.25\%$ and $1.14 \pm 0.58\%$, respectively. This indicates that while similar quantities were recovered by the 'light-petroleum method' and the reference method [20], a significantly greater quantity was recovered by MAE.

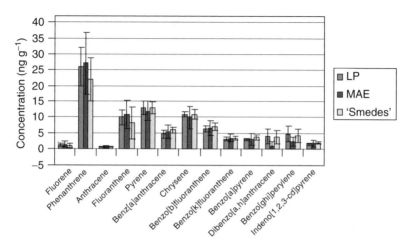

Figure 5.19 Comparison of extraction methods for the recovery of PAHs from earth-worms using reversed-phase high performance liquid chromatography–flame ionization detection (RPHPLC–FID) ($n = 15$ for each method; error bars are based on one standard deviation (SD)) [19].

5.3.1 Chemical-Extraction Methods Used to Estimate the Bioavailability of POPs by Earthworms

Approaches to assess the bioavailability of POPs in soil using non-exhaustive extraction employing organic solvents have been discussed above (Section 5.2). However, most approaches that were considered lacked any testing of correlations with bioassays, e.g. earthworms. This section proposes to review some of the work carried out in identifying whether links exist between the uptake by earthworms of POPs and their non-exhaustive extraction from soils.

Tang *et al.* [21] evaluated the use of solid-phase extraction (SPE) with C18 membrane discs and liquid-phase extraction with an 'aqueous-based' solution of tetrahydrofuran (THF) for estimating the bioavailability to earthworms of DDT, DDD and DDE in soil. The concentration of DDT and its metabolites in soil (unaged and aged) are shown in Table 5.16, while the percentage uptake of these compounds by the earthworm, *Eisenia foetida*, are also displayed in this table.

5.3.1.1 Solid-Phase Extraction

To a 1 g soil sample was added 50 ml of deionized water and a conditioned SPE disc. After a suitable duration of shaking, the discs were removed and rinsed with water to remove soil particulates. The SPE discs were then dried on paper towels and transferred to 20 ml vials. Compounds were eluted by using 5 or 10 ml of hexane.

Table 5.16 Concentration and uptake of DDT, DDD and DDE in soil and earthworms [21]

Soil	Aging time	Concentration in soil (mg kg^{-1})[a]			Uptake by *Eisenia foetida* (%)[b]		
		DDT	DDE	DDD	DDT	DDE	DDD
Sassafras silt loam	49 years	3.12 ± 0.21	3.09 ± 0.14	1.14 ± 0.11	27.8 ± 3.4	24.6 ± 2.7	30.4 ± 3.1
Chester loam	49 years	12.2 ± 1.6	4.54 ± 0.32	3.89 ± 0.23	3.53 ± 0.26	18.4 ± 1.8	6.68 ± 0.48
Dahlgren sandy loam	ca. 30 years	58.7 ± 4.7	49.3 ± 4.0	60.3 ± 5.5	3.15 ± 0.34	5.06 ± 0.61	4.82 ± 0.36
Collamer silt loam	924 days	194 ± 13	6.07 ± 0.45	3.80 ± 0.32	9.47 ± 1.28	25.4 ± 2.2	32.3 ± 3.0
Lima loam no. 1	186 days	22.9 ± 3.2	0.58 ± 0.04	0.65 ± 0.08	19.4 ± 1.6	60.3 ± 3.4	66.2 ± 5.5
Lima loam no. 2	0 days	209 ± 17	1.96 ± 0.11	3.30 ± 0.35	17.7 ± 1.8	43.8 ± 5.1	56.7 ± 6.2
Hudson silt loam	924 days	78.5 ± 6.3	3.12 ± 0.20	1.12 ± 0.16	13.5 ± 1.5	24.3 ± 2.1	27.7 ± 2.4
Hudson silt loam	0 days	155 ± 13	1.50 ± 0.08	2.40 ± 0.18	17.6 ± 1.6	54.8 ± 4.8	53.6 ± 5.1
Kendaia loam	186 days	197 ± 17	4.03 ± 0.28	19.7 ± 2.3	10.8 ± 1.1	20.5 ± 2.1	17.4 ± 1.4
Kendaia loam	0 days	248 ± 17	2.12 ± 0.13	3.58 ± 0.31	13.6 ± 1.4	45.8 ± 3.4	53.0 ± 4.5

[a]Triplicate soil samples (1 g dry weight) extracted by using a Soxhlet apparatus (10–12 h with 120 ml of hexane); extracts concentrated and made up to 5 or 10 ml in hexane, followed by analysis by gas chromatography–electron-capture detection (GC–ECD).
[b]Triplicate worm samples (previously frozen at −10°C) ground with 25–30 g of anhydrous Na$_2$SO$_4$, prior to Soxhlet extraction (as for soil). Data presented as percentages of compounds in soil that were assimilated by the earthworms.

5.3.1.2 Solvent Extraction

To a 1 g soil sample was added 20 ml of a 25% aqueous solution of THF. Samples were then shaken for 24 h on a reciprocating shaker. The sample was next centrifuged at 15 000 g for 20 min. Compounds were recovered by using a C18 extraction membrane which resulted in the extract being prepared in 5 or 10 ml of hexane.

Statistical analysis of the data using linear regression analysis indicated that the SPE approach gave data that were 'well-correlated' with earthworm uptake.

SAQ 5.8

What is the form of the equation for linear regression analysis?

The correlation coefficients were 0.967, 0.984 and 0.940 for DDT, DDE and DDD, respectively. The corresponding equations for the regression lines were $y = 10.69x - 1.01$, $y = 6.54x + 1.59$ and $y = 10.81x + 1.13$. Similarly, for the solvent-extraction approach, using an aqueous solution of THF, the data were 'well-correlated' with earthworm uptake. The correlation coefficients were 0.918, 0.973 and 0.831 for DDT, DDE and DDD, respectively. The corresponding equations for the regression lines were $y = 2.77x - 0.16$, $y = 2.20x + 1.72$ and $y = 2.11x + 2.42$. Both approaches therefore demonstrate promise as chemical-extraction methods for predicting the bioavailability of DDT and its metabolites. The same workers [22] also evaluated the uptake of DDT and its metabolites (as well as dieldrin) by the earthworm *Eisenia foetida* from a site that had been treated with DDT 49 years previously. As well as measuring the uptake of the compounds by earthworms, they also investigated the use of solid-phase extraction with Tenax TA (a porous polymer based on 2,6-diphenyl-*p*-phenylene oxide). The results, however, were not as good as those reported previously using C18 SPE membrane discs [21].

Using the same approaches of solid-phase extraction (SPE) with C18 membrane discs and liquid-phase extraction with either an aqueous-based solution of tetrahydrofuran (THF) or 95% ethanol, Tang *et al.* [23] also estimated the bioavailability to the earthworm *Eisena fetida* of polycyclic (polynuclear) aromatic hydrocarbons (PAHs). Studies were conducted with four PAHs, namely anthracene, chrysene, pyrene and benzo(a)pyrene, from five different soils and with pyrene only from six different soils. Bioavailability was assessed by uptake by the earthworms. These authors [23] found that the uptake by earthworms of the PAH mixtures when correlated with THF extraction were highly correlated ($R^2 > 0.85$);¶ slightly inferior correlations were noted when the PAHs

---

¶ R is known as the *correlation coefficient* and provides a measure of the quality of calibration. In fact, R^2 (the *coefficient of determination*) is often used because it is more sensitive to 'changes'. This varies between -1 and $+1$, with values very close to -1 and $+1$ pointing to a very tight 'fit' of the corresponding calibration curve.

were extracted by C18 membrane discs ($R^2 > 0.77$). Correlations (R^2) for the 'pyrene-only' work were 0.71 and 0.82 for THF extraction and C18 membrane discs, respectively. Data for extraction with ethanol were variable.

Other workers have, for instance, evaluated the effectiveness of different solvent-extraction systems to assess the bioavailability of POPs. For example, Yu *et al.* [24] evaluated six different solvent systems for predicting the bioavailability of butachlor and myclobutanil from soil by using the earthworms, *Eisenia foetida* and *Allolobophera caliginosa*. Extractions were carried out by using the following solvent systems: methanol, methanol–water (9:1 (vol/vol)), methanol–water (1:1 (vol/vol)), acetone–water (5:3 (vol/vol)), petroleum ether (60–90°C) and water. Soil samples (10 g) were extracted with 50 ml of solvent(s) by shaking on a reciprocating shaker for 2 h at $20 \pm 2°C$. Extracts were then filtered prior to concentration. In addition, and by way of comparison, soil samples (10 g) were 'Soxhlet-extracted' with methanol for 24 h and then concentrated. The results for the recovery of butachlor from soil using different solvent systems are shown in Figure 5.20.

DQ 5.16

What effect does aging have on the recovery of butachlor?

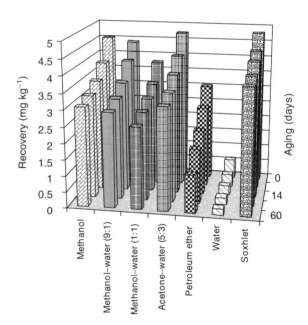

Figure 5.20 Recovery of butachlor from soil ($n = 3$) after aging and with different solvent systems [24].

Answer

It is noted that aging hinders the recovery of butachlor.

Maximum recoveries were obtained by using Soxhlet-extraction. Similar results were obtained for myclobutanil. Earthworms were exposed to soil samples after suitable aging (0, 7, 14, 30 and 60 days). Earthworms (ten mature in the case of *E. foetida* and three mature for *A. caliginosa*) were introduced onto the surface of the soil. After 7 days exposure, the earthworms were removed from the soil and kept for 24 h on moistened filter paper to allow 'purging' of the gut contents. Earthworms were then frozen prior to Soxhlet-extraction. The earthworms samples (2.0–2.5 g) were ground with 10 g of anhydrous sodium sulfate and sequentially extracted with 80 ml of dichloromethane and 80 ml of ethyl acetate for 12 h, respectively. Extracts were then concentrated and 'cleaned-up' by using deactivated aluminium oxide prior to analysis by GC. The results for butachlor are shown in Figure 5.21 (similar results were also obtained for myclobutanil).

DQ 5.17

What effect does the uptake by earthworms have on the recovery of butachlor from aged soils?

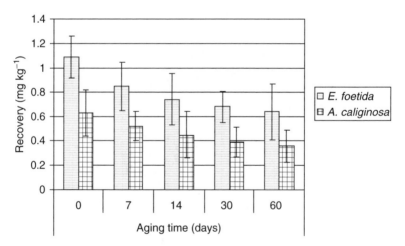

Figure 5.21 Recovery of butachlor from earthworm tissues after soil aging (*n* = 3; error bars represent one standard deviation (SD) [24].

Answer

It is noted that the uptake of butachlor by earthworms decreased with soil aging.

Statistical analysis of the data by using linear regression analysis indicated excellent correlation between the earthworm uptake of butachlor and extraction with water. The equation for *E. foetida* was $y = 8.477x + 1.113$ ($R^2 = 0.99$, $p = 0.0001$), while for *A. caliginosa* the equation was $y = 4.931x + 0.661$ ($R^2 = 0.97$, $p = 0.002$). Similarly, for myclobutanil the data were as follows for *E. foetida*: $y = 9.257x + 1.165$ ($R^2 = 0.96$, $p = 0.004$), while for *A. caliginosa* the equation was $y = 7.125x + 1.045$ ($R^2 = 0.96$, $p = 0.004$). Good correlation was also obtained for the other solvent systems used when compared to uptake by the earthworms, with R^2 values ranging from 0.91 to 0.99. These workers indicated that the use of the solvent systems studied showed promise as chemical-extraction methods for predicting bioavailability.

5.4 Plant Uptake

A recent review has highlighted the procedures required for the preparation of plant material for subsequent analysis by chromatography [25]. The sample preparation stages were identified as follows (not all are required for a given plant analysis):

- Pre-washing

- Drying plant material or freeze-drying

- Grinding

- Digestion

- Isolation and pre-concentration

- Decanting and centrifuging

- Drying the extract

- Solvent evaporation

- Extract 'clean-up'.

The crucial step of sample preparation was identified as selective 'isolation' and enrichment. The major extraction techniques (see also Chapter 3, Section 3.2) used for plant samples are:

- Conventional liquid extraction
- Soxhlet extraction
- Sonication
- Accelerated solvent extraction
- Supercritical fluid extraction.

For an introduction to uptake by plants of contaminants, including phytoremediation and bioavailability studies, see Chapter 4, Section 4.5. The uptake of non-ionic organic chemicals by plants has recently been reviewed [26]. In this review, the authors identify the major pathways for plant uptake as follows:

- **Passive and active uptake from soil into plant roots**. POP uptake has largely been identified as a passive, diffusive process by plant roots from either the vapour or water phases of soil.

- **Transport from the root to the shoot**. The major transport route through a plant is the *xylem*. However, for POPs taken up into the roots to reach the xylem requires them to cross a number of boundaries, including the epidermis, cortex, endodermis and pericycle. In the endodermis, all POPs must pass through at least one cell membrane. The key to the transport of POPs from the roots to the shoots lies in their solubilities in the water and lipid phases (cell membrane). It is, therefore, common to find that models to predict the transport of POPs within plants use the octanol–water partition coefficient (K_{ow}). This partition coefficient is used to assess the solubility of a POP in either water or octanol.

DQ 5.18

What is the octanol in the octanol–water partition coefficient mimicking?

Answer

The octanol phase is mimicking the lipid component of, for example, the cell membrane.

- **Transfer from soil to 'soil water' and 'soil air'**. POPs are sorbed onto the organic component of the soil. Therefore, to assess the transfer of POPs requires the use of another partition distribution coefficient, K_d. This coefficient is defined as the amount of POP distributed between the soil and soil water. K_d can be defined in terms of the soil organic carbon content as follows: $K_d = K_{oc} \times f_{oc}$, where K_{oc} is the organic carbon-to-water partition coefficient and f_{oc} is the fraction of organic carbon. The most widely used link between K_{oc} and K_{ow} is the Karickhoff equation [27], of which the general form for all compound classes is as follows:

$$\log K_{oc} = (0.989 \times \log K_{ow}) - 0.346 f_{oc} \qquad (5.10)$$

Any increase in f_{oc} therefore reduces the total amount of POP absorbed by the plant and the optimum K_{ow} for plant uptake.

- **Gaseous uptake from air**. POPs can volatilize from soil into the air and then diffuse into plant leaves via the cuticle or stomata (main entry route for volatile organic compounds).

- **Particulate deposition to plant surfaces**. POPs present in soil can become airborne and then re-deposited onto plant foliage via wet (e.g. rain, dew) or dry (e.g. wind) deposition.

- **Influence of plant composition on uptake**. The main aspect of plants that governs this uptake pathway is believed to be their lipid content.

In addition, other potentially significant processes include the following:

- **Plant metabolism**. POPs can become metabolized within a plant, thereby reducing their concentrations. Important aspects of this process are:

  - Transport (as described above)

  - Transformation reactions that include oxidation

  - Conjugation with, for example, glutathione

  - Sequestration into, for example, the cell wall.

- **Photolytic degradation on plant surfaces**. The direct influence of sunlight acting on POPs present on leaves.

- **Volatilization from foliage to air**. Loss of POPs from leaves. Process likely to be important for POPs with high water solubilities and high vapour pressures.

- **Growth dilution**. As the plant grows, the overall concentration of POP will reduce.

A number of recent articles highlight research that is being undertaken on the uptake of POPs by plants. A summary is given in Table 5.17.

Table 5.17 Examples of recent research investigating the uptake of POPs by plants

POP[a]	Plant type	Comments[a,b]	Extraction/Analysis[a]	Reference
Aldicarb	Corn, mung bean and cowpea	Study of the degradation of aldicarb, an oxime carbamante insecticide, in soils and uptake by plants	Samples extracted by LSE with acetone–water (3:1 (vol/vol)). Analysis by GC–FPD	Sun *et al.* [28]
Chlordane	Lettuce, pumpkin, zucchini, cucumber, tomato, thistle, lupin and spinach	Potential to phytoremediate contaminated sites	Analysis by GC–MS	Mattina *et al.* [29]
DDT, DDD and DDE	Wheat	An approach to develop a sequential-extraction approach to evaluate bioavailability of DDT and its metabolites	Samples extracted by ASE with sequential extraction with water, *n*-hexane and *n*-hexane–acetone (1:1 (vol/vol)). Analysis by GC–ECD	Tao *et al.* [30]
PAHs	Poplar trees	A 7-year study to evaluate effectiveness of hybrid poplar trees to remediate PAHs in soil and groundwater from a creosote-contaminated site	Groundwater samples extracted using LLE with dichloromethane; soil samples extracted by LSE with dichloromethane. Analysis by GC–FID	Widdowson *et al.* [31]
PAHs	White pine (*Pinus strobes*) needles	Significant variation in PAH content of pine needles in urban samples over a 7-month period; no such variation in mountain samples over the same time-period	Samples extracted by SFE using toluene-modified CO_2. Analysis by GC–MS	Lang *et al.* [32]

PAHs	Vegetables (cabbage, carrot, lettuce, leek, endive)	PAH solubility and octanol–water partition coefficient were good predictors of PAH accumulation in inner roots while vapour pressures and octanol–air partition coefficients were good predictors of PAH accumulation in leaf tissue	Vegetable samples Soxhlet-extracted with dichloromethane; soil samples Soxhlet-extracted with cyclohexane. Analysis by HPLC–FD	Kipopoulou *et al.* [33]
PAHs, PCBs and OCPs	Vegetables	POP residues determined in four varieties of organically produced potatoes and three varieties of organically produced carrots investigated	Soxhlet-extraction using hexane–acetone (1:1 (vol/vol)). Analysis by GC–MS	Zohair *et al.* [34]
Phenanthrene and pyrene	Plants (amaranth, flowering Chinese cabbage, radish, water spinach, green soybean, kidney bean, pakchoi, broccoli capsicum, eggplant and ryegrass)	Plant uptake and accumulation was correlated with soil concentration	Sonication-extraction using hexane–acetone (1:1 (vol/vol)). Analysis by HPLC–UV	Gao and Zhu [35]
OCPs	Vegetables	Evaluation of POP concentration in vegetables grown on an organic farm and conventionally determined	Soxhlet-extraction using hexane–dichloromethane (1:1 (vol/vol)). Analysis by GC–ECD	Gonzalez *et al.* [36]

(continued overleaf)

Table 5.17 (*continued*)

POP[a]	Plant type	Comments[a,b]	Extraction/Analysis[a]	Reference
OCPs (21)	Vegetation	Study of vegetation in a highly contaminated site	MAE using hexane–acetone (1:1 (vol/vol)). Analysis by GC–ECD and/or GC–MS	Barriada-Pereira *et al.* [37]
OCPs (21)	Vegetation (grass and wild plants)	An assessment of extraction techniques to recover OCPs from plants. Both MAE and Soxhlet-extraction were suitable	MAE versus Soxhlet-extraction using hexane–acetone (1:1 (vol/vol)). Analysis by GC–ECD	Barriada-Pereira *et al.* [38]
OCPs	Leek (*Allium porrum*)	POP concentration in leeks grown on an organic farm determined. Leeks bioaccumulate OCPs efficiently in their aerial and root tissues and alter the soil concentration where they are grown	Soxhlet-extraction using hexane–dichloromethane (1:1 (vol/vol)). Analysis by GC–ECD	Gonzalez *et al.* [39]
OCPs	Tomato (*Lycopersicon esculentum*)	POP concentration in tomatoes grown on an organic farm determined. Tomatoes accumulate OCPs from the soil	Soxhlet-extraction using hexane–dichloromethane (1:1 (vol/vol)). Analysis by GC–ECD	Gonzalez *et al.* [40]

[a] PAH, polycyclic (polynuclear) aromatic hydrocarbon: PCB, polychlorinated biphenyl: OCP, organochlorine pesticide.
[b] ASE, accelerated solvent extraction: GC–ECD, gas chromatography with electron-capture detection: GC–FID, gas chromatography with flame-ionization detection: GC–FPD, gas chromatography with flame-photometric detection: GC–MS, gas chromatography–mass spectrometry: HPLC–FD, high performance liquid chromatography with fluorescence detection: HPLC–UV, high performance liquid chromatography with ultraviolet detection: LLE, liquid–liquid extraction: LSE, liquid–solid extraction: SFE, supercritical fluid extraction: MAE, microwave-assisted extraction.

Summary

The different sample preparation techniques for non-exhaustive extraction (cyclodextrin, supercritical fluid extraction, sub-critical water extraction, solid-phase microextraction and membrane separations) of persistent organic pollutants from soils and sediments are described. In addition, a mathematical approach to predict weak and strong solvents, based on the Hildebrand solubility parameter, is proposed. The use of bioassays, in the form of earthworms and plants, for assessing persistent organic pollutant bioavailability is reviewed.

References

1. Lide, D. R. (Ed.), *CRC Handbook of Chemistry and Physics*, 73rd Edition, CRC Press, Boca Raton, FL, USA, 1992–1993.
2. Hatzinger, P. B. and Alexander, M., *Environ. Sci. Technol.*, **29**, 537–545 (1995).
3. Kelsey, J. W., Kottler B. D. and Alexander, M., *Environ. Sci. Technol.*, **31**, 214–217 (1997).
4. Liste, H.-H. and Alexander, M., *Chemosphere*, **46**, 1011–1017 (2002).
5. Macleod, C. J. A. and Semple, K. T., *Soil Biol. Biochem.*, **35**, 1443–1450 (2003).
6. Barton, A. F. M., *The Handbook of Solubility Parameters and other Cohesion Parameters*, CRC Press, Boca Raton, FL, USA, 1983.
7. van Krevelen, D. W. and Hoftzyer, P. J., *Properties of Polymers: Their Estimation and Correlation with Chemical Structure*, Elsevier, Amsterdam, The Netherlands, 1976.
8. Fitzpatrick, L. J. and Dean, J. R., *Anal. Chem.*, **74**, 74–79 (2002).
9. Scott, W. C. and Dean, J. R., *J. Environ. Monit.*, **5**, 724–731 (2003).
10. Szejtli, J., *Chem. Rev.*, **98**, 1743–1753 (1998).
11. Wang, X. and Brusseau, M.L., *Environ. Sci. Technol.*, **29**, 2346–2351 (1995).
12. Reid, B. J., Stokes, J. D., Jones, K. C. and Semple, K. T., *Environ. Sci. Technol.*, **34**, 3174–3179 (2000).
13. Swindell, A. L. and Reid, B. J., *Chemosphere*, **62**, 1126–1134 (2006).
14. Dean, J. R., *Methods for Environmental Trace Analysis*, AnTS Series, John Wiley & Sons, Ltd, Chichester, UK, 2003.
15. Bjorklund, E., Bowadt, S., Mathiasson, L. and Hawthorne, S. B., *Environ. Sci. Technol.*, **33**, 2193–2203 (1999).
16. Nilsson, T., Sporring, S. and Bjorklund, E., *Chemosphere*, **53**, 1049–1052 (2003).
17. Hawthorne, S. B. and Grabanski, C. B., *Environ. Sci. Technol.*, **34**, 4103–4110 (2000).
18. Hawthorne, S. B., Trembley, S., Moniot, C. L., Grabanski, C. B. and Miller, D. J., *J. Chromatogr., A*, **886**, 237–244 (2000).
19. Mooibroek, D., Hoogerbrugge, R., Stoffelsen, B. H. G., Dijkman, E. M, Berkhoff, C. J. and Hogendoor, E. A., *J. Chromatogr., A*, **975**, 165–173 (2002).
20. Smedes, F., *Analyst*, **124**, 1711–1718 (1999).
21. Tang, J., Robertson, B. K. and Alexander, M., *Environ. Sci. Technol.*, **33**, 4346–4351 (1999).
22. Morrison, D. E., Robertson, B. K. and Alexander, M., *Environ. Sci. Technol.*, **34**, 709–713 (2000).
23. Tang, J., Liste, H.-H. and Alexander, M., *Chemosphere*, **48**, 35–42 (2002).
24. Yu, Y. L., Wu, X. M., Li, S. N., Fang, H., Tan, Y. J. and Yu, J. Q., *Chemosphere*, **59**, 961–967 (2005).
25. Zygmunt, B. and Namiesnik, J., *J. Chromatogr. Sci.*, **41**, 109–116 (2003).
26. Collins, C., Fryer, M. and Grosso, A., *Environ. Sci. Technol.*, **40**, 45–52 (2006).
27. Karickhoff, S. W., *Chemosphere*, **10**, 833–846 (1981).
28. Sun, H., Xu, J., Yang, S., Liu, G. and Dai, S., *Chemosphere*, **54**, 569–574 (2004).
29. Mattina, M. J. I., Lannucci-Berger, W., Musante, C. and White, J. C., *Environ. Poll.*, **124**, 375–378 (2003).

30. Tao, S., Guo, L. Q., Wang, X. J., Liu, W. X., Ju, T. Z., Dawson, R. Cao, J., Xu, F. L. and Li, B. G., *Sci. Total Environ.*, **320**, 1–9 (2004).
31. Widdowson, M. A., Shearer, S., Andersen, R. G. and Novak, J. T., *Environ. Sci. Technol.*, **39**, 1598–1605 (2005).
32. Lang, Q., Hunt, F. and Wai, C. M., *J. Environ. Monit.*, **2**, 639–644 (2000).
33. Kipopoulou, A. M., Manoli, E. and Samara, C., *Environ. Poll.*, **106**, 369–380 (1999).
34. Zohair, A., Salim, A-B., Soyibo, A.A. and Beck, A.J., *Chemosphere*, **63**, 541–553 (2006).
35. Gao, Y. and Zhu, L., *Chemosphere*, **55**, 1169–1178 (2004).
36. Gonzalez, M., Miglioranza, K. S. B., Aizpun de Moreno, J. and Moreno, V. *J., Food Chem. Toxicol.*, **43**, 261–269 (2005).
37. Barriada-Pereira, M., Gonzalez-Castro, M. J., Muniategui-Lorenzo, S., Lopez-Mahia, P., Prada-Rodriguez, D. and Fernandex-Fernandez, E., *Chemosphere*, **58**, 1571–1578 (2005).
38. Barriada-Pereira, M., Concha-Grana, E., Gonzalez-Castro, M. J., Muniategui-Lorenzo, S., Lopez-Mahia, P., Prada-Rodriguez, D. and Fernandex-Fernandez, E., *J. Chromatogr., A*, **1008**, 115–122 (2003).
39. Gonzalez, M., Miglioranza, K. S. B., Aizpun de Moreno, J. and Moreno, V. J., *J. Agric. Food Chem.*, **51**, 5024–5029 (2003).
40. Gonzalez, M., Miglioranza, K. S. B., Aizpun de Moreno, J. and Moreno, V. J., *J. Agric. Food Chem.*, **51**, 1353–1539 (2003).
41. Dean, J. R., *Extraction Methods for Environmental Analysis*, John Wiley & Sons, Ltd, Chichester, UK, 1998.

Chapter 6

Methods Used to Assess Oral Bioaccessibility

Learning Objectives

- To appreciate the role of oral bioaccessibility in assessing metal and POP risks to humans.
- To understand the essentials of physiology as related to the human digestive system.
- To be aware of the processes that occur within the human gastrointestinal tract.
- To understand the main aspects of *in vitro* gastrointestinal methods.
- To be aware of the main variables in *in vitro* gastrointestinal methods.
- To appreciate the different approaches that have been taken in the use of oral bioaccessibility methods for metals in environmental samples.
- To appreciate the different approaches that have been taken in the use of oral bioaccessibility methods for persistent organic pollutants in environmental samples.
- To be aware of developments in the validation of oral bioaccessibility methods.

6.1 Introduction

One of the more important exposure routes to humans and animals from environmental contaminants (i.e. metals and persistent organic pollutants) is via oral ingestion.

Bioavailability, Bioaccessibility and Mobility of Environmental Contaminants　John R. Dean
© 2007 John Wiley & Sons, Ltd

DQ 6.1

What other exposure routes can you think of?

Answer

Other exposure routes include dermal absorption and breathing contaminated air.

This chapter focuses on the approaches used to mimic *in vitro* absorption of metals and POPs from foodstuffs and soil matrices. The inclusion of soil may seem a strange choice of material to include as a product that is consumed, but this is not the case. Soil consumption can occur either intentionally or unintentionally. In the case of the former, it can occur as a result of religious belief, medicinal purposes, or as part of a regular diet [1]. In this situation, the intentional consumption of soil (or chalk) is termed *geophagy*.

SAQ 6.1

How might the unintentional consumption of soil arise?

The presence of metals and POPs in vegetable samples may result from uptake, by the plant, from the soil (or perhaps in the case of metals by contamination during processing). It is also worth remembering that the normal consumption of food by humans (and animals) results in exposure to essential elements, e.g. calcium, selenium, etc., required for normal metabolic activities as well as non-essential elements, e.g. lead, that can be toxic. Under normal circumstances, it is not expected that humans would be exposed to significant levels of POPs.

DQ 6.2

Who monitors food safety in the UK?

Answer

In the UK, the Food Standards Agency regularly monitors selected foodstuffs for pesticides and other organic pollutants to ensure that maximum residue levels remain within guidelines (typically $0.01-5\,mg\,kg^{-1}$, depending upon the pesticide).

It is therefore important to be able to assess the amount of metal or POP that is potentially available for absorption in the stomach and/or intestines, i.e. bioaccessible, or to be excreted. This chapter discusses the current methods used to study oral bioaccessibility as part of an overall estimation into assessing the chemical risk to humans of consuming a range of sample types. The approaches available to assess (oral) bioaccessibility are often described by using the terminology of an *in vitro* (simulated) gastrointestinal extraction or the physiologically based extraction test (PBET).

6.2 Introduction to Human Physiology

It is worthwhile considering the aspects of human physiology which relate directly to *in vitro* gastrointestinal extraction. A diagram of the human digestive system and associated accessory organs is shown in Figure 6.1. The digestive tract of humans is two-ended, i.e. food enters via the mouth and exits via the anus. It is possible to identify seven specialist functions of the gastrointestinal tract.

SAQ 6.2

What are the seven specialist functions of the gastrointestinal tract?

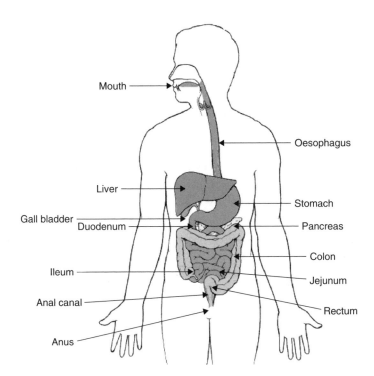

Figure 6.1 The human digestive system including the gastrointestinal tract and accessory digestive organs. Adapted from http://www.sd43.bc.ca/etip/wq_sc_life_science/gr4_sc_blank_digest.html (accessed 8 March 2006).

A simplified perspective on the processes that occur within the gastrointestinal tract are now described:

- *Mouth.* Initial breakdown of the food in the presence of saliva.

- *Oesophagus.* The oesophagus is approximately 25 cm long. By the action of wavelike muscular contractions (peristalsis), swallowed food is transferred from the mouth to the stomach.

- *Stomach.* The stomach is essentially a 'J-shaped' pouch that stores foodstuffs, initiates digestion of proteins and transports food into the small intestine or duodenum. Cells within the stomach wall secrete the inactive enzyme, pepsinogen, and hydrochloric acid (pH <2). The presence of hydrochloric acid has three functions: (1) proteins become denatured, (2) the enzyme pepsin is formed from pepsinogen and (3) pepsin is more active at pH 2. The action of pepsin within the stomach can partially digest proteins (note that carbohydrates and fats are not digested in the stomach).

- *Small intestine.* The small intestine is approximately 3 m long in a living person. It is composed of three parts: the duodenum (20–30 cm long), the jejunum (approximately 110 cm long) and the ileum (approximately 165 cm long). These parts of the small intestine contain various enzymes, including enterokinase that activates the protein-digesting enzyme, trypsin, which is secreted by the pancreas. The small intestine is the main site within the gastrointestinal tract where food products are absorbed, including fats, carbohydrates, proteins, calcium, iron, vitamins, water and electrolytes.

- *Liver and gallbladder.* As well as having a regulatory effect on the chemical composition of blood, the liver produces and secretes bile which is initially stored in the gall bladder prior to its release into the duodenum. Bile is composed of bile salts (principally cholic acid and chenodeoxycholic acid), bile pigments (bilirubin), phospholipids (lecithin), cholesterol and inorganic ions (sodium, potassium, chloride and bicarbonate).

- *Pancreas.* The pancreas secretes pancreatic juice into the duodenum. Pancreatic juice contains water, bicarbonate and a range of digestive enzymes, including amylase (digests starch), trypsin (digests protein) and lipase (digests triglyclerides).

- *Large intestine or colon.* Water and electrolytes are absorbed from the food material in the colon.

- *Defecation.* After absorption of water and electrolytes, the waste material that is left passes to the rectum. The build-up of rectal pressure, from the waste material, leads to the urge to defecate. At this point, the waste products pass through the anal canal and exit the anus.

6.3 Considerations in the Design and Development of a Simulated *in vitro* Gastrointestinal Extraction Method

6.3.1 Design of an in vitro *Gastrointestinal Method*

In vitro extraction methods are seeking to mimic the major processes that occur in the gastrointestinal tract in order to assess oral bioaccessibility. It is therefore possible to identify up to three distinct, but linked, areas of the human digestive system that are important when designing the extraction process (Figure 6.2).

DQ 6.3

What do you think the three areas are?

Answer

The three areas are the mouth, stomach and (small) intestine(s).

6.3.1.1 Mouth

The mouth is the point where the process of mechanical grinding of foodstuffs takes place at a pH of 6.5. The foodstuffs are masticated, allowing larger components to be broken down into smaller fragments, thereby increasing the surface area.

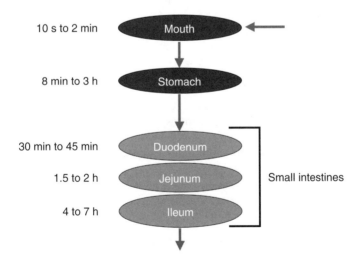

Figure 6.2 Physiology of human contaminant uptake.

DQ 6.4

What additional process occurs during mastication?

Answer
Mastication also acts to lubricate the material by the addition of saliva.

The entire process in the mouth is relatively short (a few seconds to minutes). The samples are then transported, via the oesophagus, into the stomach in a few seconds.

6.3.1.2 Stomach

The hydrochloric acidic environment of the stomach (pH 1–5) will allow dissolution of labile mineral oxides, sulfides and carbonates, thereby releasing metals. The presence of pepsin will act to break down proteins, thereby aiding dissolution of the foodstuffs. The entire process in the stomach can last from a few minutes (typically 8 min) to several hours (3 h).

6.3.1.3 Small Intestine(s) (Duodenum, Jejunum and Ileum)

Samples are subjected to digestion in the small intestine(s) by intestinal juices composed of enzymes (trypsin, pancreatin and amylase), bile salts and bicarbonate. The breakdown of food materials in the small intestine(s) means that the components are more amenable to absorption. Samples can be retained in the small intestines as follows: in the duodenum for between 30 and 45 min (at pH 4–5.5), in the jejunum for 1.5 to 2 h (at pH 5.5–7.0) and in the ileum for 5 to 7 h (at pH 7.0–7.5). The absorption process mostly takes place in the small intestine where final digestion occurs.

6.3.2 Development of an in vitro Gastrointestinal Method

Based on the above discussions, the development of an *in vitro* gastrointestinal method should have the following features:

- *Temperature*. All extractions should take place at body temperature, i.e. 37°C.

- *Shaking* or *agitation*.

SAQ 6.3

How can the peristaltic actions of the oesophagus be mimicked?

- *Mouth compartment (optional)*. Due to the short time-period that food may spend in the mouth, this aspect may not be as important as that concerning the stomach and small intestine(s). If included, the sample should be subjected to saliva at pH 6.5 for approximately 2 min.

- *Stomach compartment.* Samples are subjected to pepsin and hydrochloric acid at a pH of 1–4 for approximately 3 h.

- *Small intestine(s) compartment.* Samples are subjected to intestinal juices at a pH of 4–7.5 for approximately 7 h.

The chemical composition of saliva, gastric juice, duodenal juice and bile juice is shown in Table 6.1. In reality, it is common practice to use a simpler combination of chemicals, including inorganic ions and enzymes. In addition, a recent article [3] has identified the major variables in existing *in vitro* gastrointestinal methods as follows:

(1) **Sample-to-solution ratio.** This varies between 1:2 and 1:150 (g ml^{-1}).

(2) **Mixing and incubation time.** In order to simulate gastrointestinal mixing, the procedures employ a variety of shaking techniques, e.g. a 'shaking' water

Table 6.1 Compositions of saliva-gastro-duodenal-bile juices [2]

Juice	Organic constituents	Concentration (mg l^{-1})	Inorganic constituents	Concentration (mg l^{-1})
Saliva	α-Amylase	145	KCl	895
	Mucin	5	KSCN	200
	Urea	200	Na$_2$SO$_4$	570
	Uric acid	15	NaCl	290
			NaH$_2$PO$_4$	885
Gastric	Glucosaminhydrochloride	330	CaCl$_2$	200
	Glucose	650	HCl	1380
	Glucoronic acid	20	KCl	820
	Mucin	1500	NaCl	2750
	N-Acetylneuraminic acid	50	NaH$_2$PO$_4$	270
	Pepsin, porcine	1000	NH$_4$Cl	305
	Serumalbumin, bovine	1000		
	Urea	85		
Duodenal	Lipase	500	CaCl$_2$	200
	Pancreatin	3000	KCl	560
	Serumalbumen, bovine	1000	KH$_2$PO$_4$	80
	Stearic acid	5	MgCl$_2$	50
	Urea	100	NaCl	7010
			NaHCO$_3$	1800
Bile	Bile, chicken	3000	CaCl$_2$	220
	Serumalbumen, bovine	1800	KCl	370
	Urea	250	NaCl	5250
			NaHCO$_3$	4200

bath, mechanical stirring, argon gas dispersion, 'end-over-end' rotation or peristaltic movement. Samples are mixed for different periods of time; for gastric juice extraction between 1 and 4 h, whereas a 1–5 h incubation is used for intestinal juice extraction.

(3) **Addition of alimentary components.** Whole milk powder or cream may be added to the synthetic gastric juices to simulate the influence of food on the mobilization of the contaminants and be more representative of human digestion.

(4) **Mouth compartment (optional stage).** As the pH of saliva is close to neutral, the inclusion of the mouth compartment in any model is not expected to facilitate significant dissolution from foodstuffs.

(5) **Gastric juice.** The major enzyme involved in gastric juice extraction consists of a solution containing pepsin in the concentration range $1.25–10 \, mg \, ml^{-1}$. The enzyme is prepared in a dilute hydrochloric acid solution with a pH that varies between 1.1 and 4.0.

(6) **Intestinal juice.** The main enzymes used in intestinal juice are pancreatin and bile salt. Pancreatin is a mixture of lipase (to dissolve fat), protease (to digest proteins) and amylase (to break down carbohydrates). The addition of bile salts acts to help dissolve fat and aid in absorption in the small intestine. The pH of intestinal juice is maintained in the range 5.0 to 8.0.

(7) **Analysis of *in vitro* extracts.** The choice of analytical technique, and any sample pre-treatment required, to measure the metal or persistent organic pollutant content from each simulated extraction step could influence the data reported, e.g. the requirement for metal–matrix-matched standards for calibration in ICP–MS (see Chapter 2) or the use of liquid–liquid extraction (or solid-phase extraction) prior to POP analysis (see Chapter 3).

6.4 Approaches to Assess the Bioaccessibility of Metals

Application of gastrointestinal extraction methods for assessing metal bioaccessibility in soil and food samples are reviewed in Table 6.2. As can be observed, a wide range of metals, such as Al, As, Cd, Ce, Cu, Fe, Hg, Mg, Pb, Se and Zn, have been studied.

DQ 6.5

Consider the different applications to which the oral bioaccessibility of metals has been applied in Table 6.2.

Table 6.2 Applications of gastrointestinal extraction methods (*in vitro*) to metal bioavailability in soil and food samples (adapted from Intawongse and Dean [3])[a]

Analyte	Sample matrix	Simulated digestive fluid	Extraction conditions plus analytical technique	Comments	Reference
Al	'Fast foods', e.g. beef-, chicken-, fish- and pork-based products	Gastric juice: 16 g pepsin in 100 ml of 0.1 M HCl. Intestinal juice: 4 g pancreatin and 25 g bile extract in 1 l of 0.1 M NaHCO$_3$	Gastric digestion: 10 g (dry weight) of homogenized sample was added to 80 ml of deionized water and the pH was adjusted to 2.0; 3 ml of gastric juice added. The mixture was made up to 100 g with deionized water and incubated in a 'shaking' water bath (37°C, 2 h). Intestinal digestion: 20 g of gastric digest was adjusted to pH 7.5, 5 ml of intestinal juice added and incubated (37°C, 2 h). Aluminium was determined by ETAAS	The bioaccessible fraction of Al estimated with *in vitro* assays was between 0.85 and 2.15%	Lopez *et al.* [4]
As	Soil	Gastric juice: 1.25 g pepsin, 0.5 g sodium citrate, 0.5 g malate, 420 µl lactic acid and 500 µl acetic acid were added to 1 l deionized water, pH 2.5. Intestinal media: 175 mg bile salt and 50 mg pancreatin	Gastric digestion: 1 g of soil sample was combined with 100 ml gastric juice, placed in a water bath (37°C) and mixed by bubbling nitrogen through the solution for 1 h. Intestinal digestion: gastric digest neutralized to pH 7.0 and intestinal media added (37°C, 4 h mixing). Arsenic was determined by ICP–AES	The results indicated an average stomach absorption of 11.2% of total soil As. The absorption increased to 18.9% in a simulated small intestine regime	Williams *et al.* [5]

(continued overleaf)

Table 6.2 (*continued*)

Analyte	Sample matrix	Simulated digestive fluid	Extraction conditions plus analytical technique	Comments	Reference
As	Contaminated soils and solid media (waste)	Gastric juice: 1% pepsin in 0.15 M NaCl. Intestinal media: 2.10 g bile extract and 0.21 g pancreatin	Gastric digestion: 4 g (air-dried) soil sample was added to 600 ml gastric solution. An equivalent amount of the dosing vehicle (200 g dough) was added to the solution, pH adjusted to 1.8 and incubated in a water bath (37°C, 1 h). Intestinal digestion: gastric-phase solution was adjusted to pH 5.5 and intestinal media added and incubated (37°C, 1 h). Arsenic was determined by HyICP–AES	Arsenic extracted by the *in vitro* gastrointestinal stomach and intestinal phases and linearly correlated with *in vivo* bioavailable arsenic ($R = 0.83$ and 0.82, respectively)	Rodriguez *et al.* [6]
As, Pb	Soil (As), mine wastes (Pb)	Gastric juice: 1.25 g pepsin, 0.5 g citrate, 0.5 g malate, 420 µl lactic acid and 500 µl acetic acid were added to 1l distilled water, pH 1.3, 2.5, 4.0. Intestinal media: 70 mg bile salt and 20 mg pancreatin	Gastric digestion: 0.4 g (air-dried) of test material was combined with 40 ml gastric juice, placed in a water bath (37°C), allowed to stand for 10 min and mixed for 1 h. Intestinal digestion: gastric digest neutralized to pH 7.0 and intestinal media added (37°C, 3 h shaking). Metals were determined by ICP–AES	For Pb, the results were linearly correlated with *in vivo* results ($R^2 = 0.93$). For As, the results were 'overpredicting' *in vivo* results (2–11% difference between *in vitro* and *in vivo* results)	Ruby *et al.* [7]
Cd	Contaminated soils	Gastric juice: 1% pepsin in 0.15 M NaCl. Intestinal media: 2.10 g bile extract and 0.21 g pancreatin	Gastric digestion: 4 g (air-dried) soil sample was added to 600 ml gastric solution. An equivalent amount of the dosing vehicle (200 g dough) was added to the solution, pH adjusted to 1.8 and incubated	Bioaccessible Cd measured by *in vitro* gastrointestinal extraction method and the results linearly correlated with *in vivo* relative bioavailable Cd ($R = 0.80$)	Schroder *et al.* [8]

Analyte	Food type	Gastric/intestinal juice	Procedure	Comments	Reference
Cd, Cu, Fe, Pb, Zn	Cereals, meats, fish and vegetables	Gastric juice: 1% pepsin in saline–HCl, pH 1.8. Intestinal media: equal volumes of 1.5% pancreatin and bile salts, in saline solution	in a water bath (37°C, 1 h). Intestinal digestion: gastric-phase solution was adjusted to pH 5.5 and intestinal media added and incubated (37°C, 1 h). Cadmium was determined by ICP–AES. Gastric digestion: 100 g (wet weight) food sample and deionized water (50 ml) was mixed in a Stomacher bag. Gastric juice was added (125 ml, in three aliquots). The digest were mixed using the Stomacher bag and incubated (37°C, 2 h). Intestinal digestion: gastric digest was adjusted to pH 7.5 and intestinal juice (125 ml) added, mixed and incubated at 37°C for 2 h. Cadmium and lead were determined by ETAAS, while copper, iron and zinc were determined by DCP–AES	The results show that the amount of soluble analyte varies widely from element to element and according to food type	Crews *et al.* [9]
Cd, Cu, Fe, Pb, Zn	Foods, e.g. beef, soya, crab, bread and wholemeal	Gastric juice: 10 mg ml^{-1} pepsin in saline–HCl (0.15 M NaCl: 0.02 M HCl) at pH 1.8. Intestinal media: equal volumes of (a) 30 mg ml^{-1} pancreatin plus 10 mg ml^{-1} amylase and (b) bile salts (1.5 g l^{-1}) in 0.15 M NaCl	Gastric digestion: 50 g (wet weight) food was mixed with 100 ml gastric juice in a Stomacher bag and incubated in a 'shaking' water bath (37°C, 4 h). Intestinal digestion: gastric digest was adjusted to pH 7.4 and 100 ml	Enzymatic action can change the nature of soluble species and increase the amount of soluble analyte. The solubility can be affected by the presence of another food	Crews *et al.* [10]

(continued overleaf)

Table 6.2 (*continued*)

Analyte	Sample matrix	Simulated digestive fluid	Extraction conditions plus analytical technique	Comments	Reference
			of intestinal juice added and incubated (37°C, 4h). Cadmium and lead were determined by ETAAS, while copper, iron and zinc were determined by DCP–AES		
Cd, Pb	Cocoa powder and liquor	Gastric juice: 50 mg pepsin in 5 ml 150 mM NaCl, pH 2.5. Intestinal media: equal volumes of 3% (wt/vol) pancreatin plus 1% (wt/vol) amylase in water and 1.5 g l⁻¹ bile salts in water	Gastric digestion: 1 g sample (powder) suspended in 5 ml of gastric solution and incubated (37°C, 4 h). Intestinal digestion: gastric digest adjusted to pH 7.4 and 5 ml of intestinal media added (37°C, 4 h shaking). Cadmium and lead were determined by ICP–MS	The bioaccessibility of Cd in cocoa varied from 10 to 50% in gastrointestinal conditions, while the bioaccessibility of Pb did not exceed 5–10%	Mounicou *et al.* [11]
Ce, Fe, Mg and trace elements	Soil	Gastric juice: 1.25 g pepsin, 0.5 g citrate, 0.5 g malate, 420 µl lactic acid and 500 µl acetic acid were added to 1 l deionized water, pH 2.5. Intestinal media: 175 mg bile salt and 50 mg pancreatin	Gastric digestion: 1 g (air-dried) of soil sample was mixed with 100 ml gastric juice and placed in a shaking water bath (37°C, 1 h). Intestinal digestion: gastric digest neutralized to pH 7.0 and intestinal media added (37°C, 3 h). Metals were determined by either ICP–AES or ICP–MS	Results for Ce, Mg and Fe indicated bioaccessibilities in the range of 1–15%, 7–33% and 0.1–2.9%, respectively	Smith *et al.* [12]
Fe	Foods, e.g. corned beef, bacon, steak, peas, cabbage, etc.	Gastric juice: 0.5% pepsin in 0.1 N HCl Intestinal media: NA	An aliquot of homogeneous sample was diluted with an equal volume of gastric juice, pH adjusted to 1.5–2 and incubated (37°C, 1.5h). Iron	In most cases, less than half the iron in the foods is released into solution. The amount released is influenced by cooking	Jacobs and Greenman [13]

	Foods	Gastric/intestinal juice	Method	Results	Reference
Fe	Foods and diets, e.g. cereals, vegetables, sugar, etc.	Gastric juice: 0.5% pepsin in 0.1 N HCl. Intestinal media: NA	2 g of dry food was mixed with 25 ml gastric juice. In the case of homogeneous diets, they were diluted with an equal volume of gastric juice, pH adjusted to 1.35 and incubated in a metabolic 'shaking' water bath (37°C, 1.5 h). One portion of gastric digest was adjusted to pH 7.5 and incubated (37°C, 1.5 h). Iron was determined by an α,α'-dipyridyl method	and the presence of iron-binding substances in some foods was determined by a tripyridyltriazine method. The % bioaccessible iron at pH 7.5 in a number of diets was shown to correlate with % iron absorption from the same diets observed in adult males. The method can be used as a reliable measure of bioavailability of iron in foods	Rao and Prabhavathi [14]
Fe	Foods, e.g. cocoa power, eggs, oats, peas, spinach, lettuce, etc.	Gastric juice: 5 g l^{-1} pepsin in 0.1 N HCl, pH adjusted to 1.4–2.8. Intestinal media: NA	0.2 g of dry foods were incubated with 5 ml gastric juice (37°C, 1.5 h). Iron was determined by AAS	The results correlated highly with the *in vivo* absorption of Fe from similar foods	Lock and Bender [15]
Fe	Test meals	Gastric juice: 16 g pepsin in 0.1 N HCl (made up to 100 ml with 0.1 N HCl). Intestinal media: 4 g pancreatin and 25 g bile extract in 0.1 M NaHCO$_3$ (made up to 1 l with 0.1 M NaHCO$_3$)	Gastric digestion: test meal was mixed with gastric juice in an amount that provided 0.5 g pepsin per 100 g of meal, incubated (37°C, 2 h). Intestinal digestion: 20 g aliquots of gastric digest were adjusted to pH 5. Intestinal media (5 ml) added and incubated (37°C, 2 h). Iron was determined by AAS	Relative availabilities determined for a series of meals containing ascorbic acid, eggs, orange juice, tea, coffee, cola or whole-wheat bread show that the method accurately reflects actual 'food-iron' availability	Miller et al. [16]
Fe, Zn	Human milk and	Gastric juice: 0.2 mg ml^{-1} pepsin in 0.1 M HCl. Intestinal	Gastric digestion: 30 ml milk sample was adjusted to pH 2.0	Zinc bioavailability from breast milk samples was	Bermejo et al. [17]

(continued overleaf)

Table 6.2 (*continued*)

Analyte	Sample matrix	Simulated digestive fluid	Extraction conditions plus analytical technique	Comments	Reference
	milk-based infant formulae	media: 0.15 mg ml^{-1} pancreatin in 0.1 M NaHCO$_3$	(corresponding to an adult) and pH 5.0 (newborn), 1 ml of gastric media added and incubated (37°C, 50 min). Intestinal digestion: 15 ml of gastric digest was neutralized to pH 7.0 and 1 ml of intestinal media added and incubated (37°C, 30 min). Iron and zinc were determined by FAAS	higher than those obtained from infant formulae at the newborn gastric pH. No differences in iron bioavailability from breast milk and infant formulae at different pH could be detected	
Hg, Se	Fish samples, i.e. swordfish, sardine and tuna	Gastric juice: 6% (wt/vol) pepsin in 0.15 M NaCl, acidified with HCl to pH 1.8. Intestinal media: 1.5% (wt/vol) pancreatin, 0.5% (wt/vol) amylase and 0.15% (wt/vol) bile salts in 0.15 M NaCl	Gastric digestion: 25 g of fish samples (dry weight) was mixed with 75 ml of gastric media, incubated in a water bath (37°C, 4 h) and shaken periodically. After 1 h, the pH was checked and adjusted to 3 with 6 M HCl. Intestinal digestion: pH raised to 6.8, then 50 ml of intestinal media added and incubated (37°C, 4 h) and shaken periodically. Mercury and selenium were determined by AFS	Se and Hg bioaccessibility varied depending on the type of fish analysed. Se solubility in the gastrointestinal supernatants was higher in swordfish and sardine (76 and 83%, respectively) than in tuna (50%). A low Hg bioaccessibility (9–17%) was found for all of the samples	Cabanero *et al.* [18]
Pb	Mine wastes	Gastric media: 50 mg pepsin and organic acids (0.5 ml acetate, 0.5 g citrate, 0.42 ml lactate and 0.5 g malate). Intestinal media: 20 mg pancreatin and 70 mg bile salt	Gastric digestion: 4 g (air-dried) mine waste suspended in 40 ml distilled water, gastric media added and placed on a 'wrist-action' shaker in a water bath (37°C, 2 h), pH level	Four mine waste samples with highly variable mineral composition resulted in only 4, 0.5, 6 and 2% bioaccessible Pb. The bioaccessible fraction	Ruby *et al.* [19]

Pb	Household dusts	Gastric media: 1 g citric acid, 1 g malic acid, 1 g glacial acetic acid, 1 ml lactic acid and 0.1 g pepsin added to 80 ml 0.1 M HCl. Intestinal media : 0.04 g pancreatin and 0.14 g bile extract	Gastric digestion: 8 g of dust was added in gastric solution, pH adjusted to 1.3, placed in 'shaking' water bath (37°C, 2 h). Intestinal digestion: gastric digest neutralized to pH 7.0 and intestinal media added (37°C, 3 h). Lead was determined by FAAS	maintained at 1.3. Intestinal digestion: gastric digest adjusted to pH 7.0, intestinal media added (37°C, 2 h shaking). Lead was determined by AAS; ETAAS for levels < 2 mg l^{-1} and FAAS for levels > 2 mg l^{-1}. The amount of dissolved Pb ranged from 26–46% of total Pb in the dust	*in vitro* from the mine waste (4 ± 0.2%) was in good agreement with the *in vivo* result (9 ± 4%) Oliver *et al.* [20]
Pb	Soil	Synthetic saliva: 145 mg α-amylase, 15 mg uric acid and 50 mg mucin. Gastric juice: 1 g BSA, 1 g pepsin and 3 g mucin. Intestinal juice: 1 g BSA, 3 g pancreatin and 0.5 g lipase. Bile solution: 1.8 g BSA and 6 g bile (chicken). NB: Inorganic and organic solutions were added to saliva, gastric, bile and intestinal juice, e.g. KCl, NaCl, CaCl₂, NaHCO₃, urea and made up to 500 ml with H₂O	Gastric digestion: 0.6 g soil (dry weight) was mixed with 9 ml synthetic saliva, pH 6.5 and rotated at 55 rpm (37°C, 5 min). The gastric juice, 13.5 ml, pH 1.07, was added and rotated at 55 rpm (37°C, 2 h). Intestinal digestion: 27 ml intestinal juice (pH 7.8) and 9 ml bile (pH 8) were added and rotated (37°C, 2 h). Lead was determined by either DP–ASV or ICP–MS	Only a negligible Pb fraction is present as free Pb²⁺, whereas lead phosphate and lead–bile complexes are important fractions	Oomen *et al.* [21, 22]

(continued overleaf)

Table 6.2 (*continued*)

Analyte	Sample matrix	Simulated digestive fluid	Extraction conditions plus analytical technique	Comments	Reference
Pb	Soil and pottery flakes	Same as above	Same as above. Lead was determined by ICP–MS	Pottery flake lead is considerably less bioaccessible ($0.3 \pm 0.2\%$) than lead in soil without pottery flakes (42–66% at the same site and 28–73% at other sites in the same town). Hence, the matrix of ingestion can affect the exposure assessment for lead	Oomen *et al.* [21, 22]

a AAS, atomic absorption spectroscopy; AFS, atomic fluorescence spectroscopy; DCP–AES, direct-current plasma–atomic emission spectroscopy; ETAAS, electrothermal atomization atomic absorption spectroscopy; FAAS, flame atomic absorption spectroscopy; HyICP–AES, hydride generation-inductively coupled plasma–atomic emission spectroscopy; ICP–AES, inductively coupled plasma–atomic emission spectroscopy; ICP–MS, inductively coupled plasma–mass spectrometry; DP–ASV, differential pulse–anodic stripping voltammetry; NA, not available.

Table 6.3 Comparison of different gastrointestinal approaches (adapted from Intawongse and Dean [3])[a,b]

Aspect	Method A	Method B	Method C	Method D	Method E	Method F	Method G	Method H
Element studied	Fe	Cd, Cu, Fe, Pb, Zn	As, Cd, Pb	As, Pb	As	As, Cd, Pb	As, Cd, Pb	As, Cd, Pb
Test sample	Test meals	50 g of foods, e.g. beef, soya, crab and bread	2.0 g of dry soil	0.4 g of dry soil/mine wastes	4 g of dry soil	10 g of dry soil	0.6 g of dry soil	10 g of dry soil
Solid:solution ratio	NA	1:2	1:50	1:100	1:150	1:2.5	1:38	1:30
Mixing	Water bath with shaker	Water bath with shaker	Agitator, 200 rpm	Passing Ar gas ($11\ \text{min}^{-1}$) through the reaction vessel	Ar gas dispersion	Mechanical stirring, 150 rpm	'End-over-end' rotation, 55 rpm	Peristaltic movement
Mouth								
• Volume of saliva	NA	NA	NA	NA	NA	NA	9.0 ml	50 ml
• Saliva pH							pH 6.5	pH 5
• Incubation time							5 min	5 min
Stomach								
• Gastric pH	pH 2	pH 1.8	pH 2.0	pH 1.3–4.0	pH 1.8	pH 4.0	pH 1.07	Initial pH 5.0, decreasing to pH 3.5, 2.5 and 2.0 after 30, 60 and 90 min, respectively

(*continued overleaf*)

Table 6.3 (*continued*)

Aspect	Method A	Method B	Method C	Method D	Method E	Method F	Method G	Method H
• Enzymes and other substances	16 g pepsin in 0.1 N HCl (made up to 100 ml with 0.1 N HCl)	10 mg ml⁻¹ pepsin in saline–hydrochloric acid (0.15 M NaCl; 0.02 M HCl)	pepsin, mucin and 50 g l⁻¹ whole milk powder (and no milk powder)	1.25 g pepsin, 0.5 g citrate, 0.5 g malate, 420 µl lactic acid and 500 µl acetic acid in 1 l distilled water	1% pepsin in 0.15 M NaCl, plus 200 g dough	15 g Nutrilon plus 16 g pectin, 8 g mucin, 5 g starch, 1 g cellobiose, 1 g glucose, 2 g protease peptone and 18 ml cream in 1 l distilled water	Gastric juice (pepsin, mucin, BSA)	Lipase, pepsin
• Volume of gastric juice	Volume provided 0.5 g pepsin per 100 g of meal	100 ml	100 ml	40 ml	600 ml	25 ml	13.5 ml	250 ml
• Incubation time	2 h	4 h	2 h	1 h	1 h	3 h	2 h	Gradual secretion of gastric content at 0.5 ml/min⁻¹
Small intestine								
• Intestinal pH	pH 5	pH 7.4	pH 7.5	pH 7.0	pH 5.5	pH 6.5	pH 7.8–8.0.	pH 6.5 (duodenum), pH 6.8 (jejunum), pH 7.2 (ileum)
• Enzymes and other substances	4 g pancreatin and 25 g bile extract in 0.1 M NaHCO₃ (made up to 1 l with 0.1 M NaHCO₃)	Equal volumes of (a) 30 mg ml⁻¹ pancreatin plus 10 mg ml⁻¹ amylase and (b) bile salts (1.5 g l⁻¹) in 0.15 M NaCl	Trypsin, pancreatin and 4.5 g l⁻¹ bile in chyme	70 mg bile salt and 20 mg pancreatin	2.10 g bile extract and 0.21 g pancreatin	12 g NaHCO₃, 4 g bovine bile and 0.9 g pancreatin in 1 l distilled water	27 ml duodenal juice (pancreatin and lipase) and 9 ml bile juice	Pancreatin and porcine

Volume of intestinal juice	5 ml	100 ml	100 ml	40 ml	600 ml	15 ml	36 ml	3 × 70 ml (3 sections: duodenum, jejunum and ileum)
Incubation time	2 h	4 h	6 h	1–3 h	1 h	5 h	2 h	Duodenal secretion of gastric content at 1 ml ml^{-1} (total digestion time, 360 min)
Digestate treatment centrifugation	NA	Centrifuged in glass centrifuge bottle.	7000 g, 10 min	2100 g, 25 min	10 000 rpm, 15 min	7000 g, 10 min	3000 g, 5 min	Dialysis
Analytical technique	AAS	ETAAS (Cd, Pb), DCP-AES (Cu, Fe, Zn)	AAS	ICP-AES	HG-ICP-AES	ICP-AES	ICP-MS	ICP-AES (Cd, Pb), HgAAS (As)
Reference	Miller et al. [16]	Crews et al. [10]	Oomen et al. [23]; Rotard et al. [24]	Ruby et al. [7]	Rodriguez et al. [6]	Oomen et al. [23]	Oomen et al. [23]; Rotard et al. [24]	Oomen et al. [23]; Minekus et al. [25]

^{a} All carried out at 37°C.

^{b} AAS, atomic absorption spectroscopy; DCP–AES, direct-current plasma–atomic emission spectroscopy; ETAAS, electrothermal atomization atomic absorption spectroscopy; HgAAS, hydride-generation atomic absorption spectroscopy; HG ICP–AES, hydride generation-inductively coupled plasma–atomic emission spectroscopy; ICP–AES, inductively coupled plasma–atomic emission spectroscopy; ICP–MS, inductively coupled plasma–mass spectrometry; NA, not available.

Answer

You will see that the different approaches that are used include different compositions of gastric and intestinal juices. This diversity of composition will obviously lead to different analytical results being obtained.

Table 6.3 provides a summary of *in vitro* approaches with various experimental designs that have been employed for evaluating the oral bioaccessibility of selected metals from soil and food samples.

DQ 6.6

Compare for yourself the different approaches used for assessing the oral bioaccessibility of metals (Table 6.3).

Answer

Table 6.3 highlights the diversity of approaches that have been used to assess metal oral bioaccessibility. Perhaps the most important variables to note are the differences in solid:solution ratio, the methods of mixing, the pH values of the gastric and intestinal juices and the compositions of these juices.

6.5 Approaches to Assess the Bioaccessibility of Persistent Organic Pollutants

Application of gastrointestinal extraction methods for assessing persistent organic pollutant bioaccessibility in soil and food samples are reviewed in Table 6.4.

DQ 6.7

Compare for yourself the different approaches used for assessing the oral bioaccesibility of persistent organic pollutants (Table 6.4).

Answer

As noted in DQ 6.5, the different approaches that are used include different compositions of gastric and intestinal juices. Such a diversity in composition will obviously lead to different analytical results being obtained.

As can be observed, a wide range of persistent organic pollutants, such as dioxins and furans, PAHs, PCBs, TPHs and pesticides, have been studied.

Table 6.4 Applications of *in vitro* gastrointestinal extraction to oral bioaccessibility assessment of POPs[a]

Compound	Simulated digestive extraction conditions plus analytical method	Reference
PCDDs/Fs (dioxins and furans)	2.0 g sample added to 20 ml saliva, with or without 7 g milk powder, and shaken at 200 rpm for 0.5 h at 37°C. 40 ml gastric juice added and pH set at 2.0 with 1.8% HCl, then shaken for 3 h. 40 ml duodenal juice added and pH set at 7.5 with NaHCO$_3$ and then shaken for 1.5 h; 20 ml bile added and shaken for 1.5 h. Centrifuged at 7000 g for 2 h at room temperature; supernatant filtrated through a 5 μm sieve. Analysis by high-resolution GC–MS	Wittsiepe *et al.* [2]; Rotard *et al.* [24]
PAHs and PCBs; PCDDs/Fs (dioxins and furans)	1 g sample suspended in 105 ml distilled water and pH set at 2.0 with 1.8% HCl; 10 mg pepsin and 350 mg mucin added with no foodstuff or with 7 g whole milk powder or 1.82 g grape seed oil; suspension shaken at 200 rpm for 2 h at 37°C. Gastric digest neutralized (pH 7) with solid NaHCO$_3$; 10 mg trypsin, 350 mg pancreatin and 350 mg bile added; suspension incubated for 6 h. Suspension centrifuged for 10 min at 7000 g and room temperature; supernatant ($\sim$ 150 ml) filtered through 5 μm sieve. Analysis by HPLC with fluorescence detector for PAHs and GC–ECD for PCBs; GC–MS for PCDDs/Fs	Wittsiepe *et al.* [2]; Hack and Selenka [26]
PCBs and lindane; benzo[a]-pyrene (PAH)	0.6–0.9 g sample added to 9 ml saliva (290 mg l^{-1} α-amylase, 100 mg l^{-1} mucin and 30 mg l^{-1} uric acid) at pH 6.5 and rotated at 60 rpm for 5 min at 37°C. 13.5 ml gastric juice (2 g l^{-1} pepsin, 6 g l^{-1} mucin and 2 g l^{-1} BSA) at pH 1, added and rotated for 2 h. 27 ml duodenal juice (6 g l^{-1} pancreatin, 2 g l^{-1} BSA and 1 g l^{-1} lipase) and 9 ml bile (12 g l^{-1} chicken bile and 3.6 g l^{-1} BSA), both at pH 8, added and rotated for 2 h. Centrifuged at 3000 g for 5 min. Analysis by GC–ECD for PCBs and lindane; HPLC with fluorescence detector for PAH	Oomen *et al.* [27–29]
TPHs	0.4 g sample added to 250 ml synthetic upper small intestinal digestive fluid at pH 6.5 containing 150 mM NaCl and 0.1/0.5/20 mM mixed-bile salts (sodium salts of glycocholate, glycochenodeoxycholate, glycodeoxycholate, glycolithocholate, glycolithocholate sulfate, taurocholate, taurochenodeoxycholate, taurodeoxycholate, taurolithocholate and taurolithocholate sulfate), with or without 30 mM mixed intestinal lipids (cholesterol monohydrate, oleic acid, monoolein, diolein and lecithin); stirred at 37°C for 4 h; centrifuged at 1100 g for 45 min at 37°C and supernatant permeated through a 0.45 μm filter. Analysis by GC–MS	Holman *et al.* [30]
PCDDs/Fs	10 g sample and 6 ml oleic acid added to 950 ml stomach fluid (2.5 g l^{-1} mucin, 1 g l^{-1} pepsin, 0.2 M glycine, 5 g l^{-1} BSA and 150 mM NaCl) at pH 1.5 and stirred for 1 h at 30 rpm and 37°C. pH adjusted to 7.2 with 10 ml of 50% NaOH; 600 mg porcine pancreatin and 4 g bovine bile added; stirred for 4 h. Centrifuged at 3000 g for 10 min	Ruby *et al.* [31]

(continued overleaf)

Table 6.4 (*continued*)

Compound	Simulated digestive extraction conditions plus analytical method	Reference
PAHs	20 g soil applied to 200 ml solution of 0.1 M KHCO$_3$ and 0.1 M NaCl; acidified to pH 1.5 with 1.3 ml of 5 M HCl; porcine pepsin added to a concentration of 10 mg l^{-1}; stirred at 150 rpm for 2 h at 37°C. Supplemented with 100 ml pancreatic juice (0.9 g l^{-1} porcine pancreatin powder, 12.5 g l^{-1} NaHCO$_3$ and 6 g l^{-1} Oxgall) at pH 6.3; stirred at 150 rpm for 5 h at 37°C. Supplemented with 100 ml microbiota suspension; stirred at 150 rpm for 18 h at 37°C. Centrifuged at 1500 g (stomach and duodenum digests) or 3000 g (colon digest) for 5 min. Analysis by GC–MS	Van de Wiele *et al.* [32]
Phenanthrene (PAH) and PCBs	1 g sample added into 9 ml saliva (4 g l^{-1} mucin, 1 g l^{-1} urea, 0.6 g l^{-1} Na$_2$HPO$_4$, 0.99 g l^{-1} CaCl$_2$.4H$_2$O, 0.4 g l^{-1} KCl and 0.4 g l^{-1} NaCl) at pH 6.5 and incubated for 5 min at 37°C. 14 ml gastric juice (3.2 g l^{-1} pepsin, 2 g l^{-1} NaCl and 0.7% HCl), at pH 3.0, added and incubated for 2 h. 27 ml duodenal juice (0.2 M NaHCO$_3$) plus 7 ml bile, both at pH 7.0, added and incubated for 2 h. Analysis by HPLC–UV for PAH; GC–ECD for PCBs	Pu *et al.* [33, 34]
PAHs	6 g sample added into 600 ml gastric solution (1.25 g l^{-1} pepsin, 8.8 g l^{-1} NaCl, 0.5 g l^{-1} citrate, 0.5 g l^{-1} malate, 0.04% lactic acid and 0.05% acetic acid), at pH 1.5, adjusted with 12 M HCl and stirred at 100 rpm and 37°C for 1 h. Gastric digest adjusted to pH 7.0 with saturated NaHCO$_3$; 1.2 g porcine bile extract and 0.36 g porcine pancreatin added and stirred for 4 h. Suspensions centrifuged at 7000 g for 10 min and supernatant filtered through 0.45 μm glass-fibre filter. Analysis by GC–MS	Tang *et al.* [35]
Pesticides, phenols and 'base-neutral' compounds	5 g sample added into 15 ml of 0.1–1% pepsin in saline (154 mM) at pH 1.8, adjusted with HCl and shaken at 100 rpm and 37°C for 3 h; centrifuged at 3000 rpm for 10 min at room temperature and supernatant filtrated. Gastric digest residue added to 5 ml of 3% pancreatin, 5 ml of 1% amylase and 5 ml of 0.15% bile salts, all in saline; pH adjusted to 7.0 with saturated NaHCO$_3$ or diluted NaOH and shaken for 3 h; centrifuged at 3000 rpm for 10 min and supernatant filtrated. Analysis by GC–MS	Scott and Dean [36]; Esteve-Turrillas *et al.* [37]

[a]PAH, polycyclic (polynuclear) aromatic hydrocarbon; PCB, polychlorinated biphenyl; TPH, total petroleum hydrogen; PCDD, polychlorinated dibenzo-*p*-dioxin; PCDF, polychlorinated dibenzofuran; BSA, bovine serum albumen; HPLC, high performance liquid chromatography; GC–MS, gas chromatography–mass spectrometry; GC–ECD, gas chromatography with electron-capture detection; UV, ultra violet (detection).

6.6 Validity for Measuring Bioaccessibility

A number of different *in vitro* test methods are available to measure the bioaccessibility of metals from soil and food samples, but the results are not generally comparable between methods, particularly when the tests are applied by different laboratories. Therefore, more quantitative data are required to check the validity and robustness of the different approaches. Certified Reference Materials (CRMs) need to be produced and validated to allow users to compare data for the different approaches, as well as providing material for quality control. The first CRM for the bioaccessibility testing of metals has been prepared in the UK by the British Geological Survey. This CRM provides bioaccessible and total-element information for a range of elements, including arsenic, and can be used to provide quality assurance in bioaccessibility testing.

In addition, further research should be performed to ensure that *in vitro* data are in agreement with *in vivo* results for a wider range of metals/POPs and samples.

Summary

The different approaches for assessing the oral bioaccessibility of metals and persistent organic pollutants have been considered. After a brief introduction to human physiology, the key components in the design of an *in vitro* gastrointestinal approach are discussed. Finally, examples of the use of *in vitro* gastrointestinal extraction for metals and persistent organic pollutants are presented.

References

1. Abrahams, P. W., 'Geophagy and the involuntary ingestion of soil,' in *Essentials of Medical Geology*, Elsevier, Amsterdam The Netherlands, 2005, Ch. 17, pp. 435–458.
2. Wittsiepe, J., Schrey, P., Hack, A., Selenka, F. and Wilhelm, M., *Int. J. Hyg. Environ. Health*, **203**, 263–273 (2001).
3. Intawongse, M. and Dean, J. R., *Trends Anal. Chem.*, **25**, 876–886 (2006).
4. Lopez, F. F., Cabrera, C., Lorenzo, M. L. and Lopez, M. C., *Sci. Total Environ.*, **300**, 69–79 (2002).
5. Williams, T. M., Rawlins, B. G., Smith, B. and Breward, N., *Environ. Geochem. Health*, **20**, 169–178 (1998).
6. Rodriguez, R. R., Basta, N. T., Casteel, S. W. and Pace, L. W., *Environ. Sci. Technol.*, **33**, 642–649 (1999).
7. Ruby, M. V., Davis, A., Schoof, R., Eberle, S. and Sellstone, C. M., *Environ. Sci. Technol.*, **30**, 422–430 (1996).
8. Schroder, J. L., Basta, N. T., Si, J., Casteel, S. W., Evans, T. and Payton, M., *Environ. Sci. Technol.*, **37**, 1365–1370 (2003).
9. Crews, H. M., Burrell, J. A. and McWeeny, D. J., *J. Sci. Food Agric.*, **34**, 997–1004 (1983).
10. Crews, H. M., Burrell, J. A. and McWeeny, D. J., *Z. Lebensm. Unters. Forsch.*, **180**, 405–410 (1985).
11. Mounicou, S., Szpunar, J., Andrey, D., Blake, C. and Lobinski, R., *Analyst*, **127**, 1638–1641 (2002).

12. Smith, B., Rawlins, B. G., Cordeiro, M. J. A. R., Hutchins, M. G., Tiberindwa, J. V., Sserunjogi, L. and Tomkins, A. M., *J. Geol. Soc.*, **157**, 885–891 (2000).
13. Jacobs, A. and Greenman, D. A., *Br. Med. J.*, **1**, 673–676 (1969).
14. Rao, B. S. N. and Prabhavathi, T., *Am. J. Clin. Nutrit.*, **31**, 169–175 (1978).
15. Lock, S. and Bender, A. E., *Br. J. Nutrit.*, **43**, 413–420 (1980).
16. Miller, D. D., Schricker, B. R., Rasmussen, R. R and Campen, D. V., *Am. J. Clin. Nutrit.*, **34**, 2248–2256 (1981).
17. Bermejo, P., Pena, E. M., Dominquez, R., Bermejo, A., Cocho, J. A. and Fraga, J. M., *Food Chem.*, **77**, 361–369 (2002).
18. Cabanero, A. I., Madrid, Y. and Camara, C., *Anal. Chim. Acta*, **526**, 51–61 (2004).
19. Ruby, M. V., Davis, A., Link, T. E., Schoof, R., Chaney, R. L., Freeman, G. B. and Bergstrom, P., *Environ. Sci. Technol.*, **27**, 2870–2877 (1993).
20. Oliver, D. P., McLaughlin, M. J., Naidu, R., Smith, L. H., Maynard, E. J. and Calder, I. C., *Environ. Sci. Technol.*, **33**, 4434–4439 (1999).
21. Oomen, A. G., Rompelberg, C. J. M., Bruil, M. A., Dobbe, C. J. G., Pereboom, D. P. K. H. and Sips, A. J. A. M., *Arch. Environ. Contam. Toxicol.*, **44**, 281–287 (2003).
22. Oomen, A. G., Tolls, J., Sips, A. J. A. M. and Hoop, M. A. G. T. V., *Arch. Environ. Contam. Toxicol.*, **44**, 107–115 (2003).
23. Oomen, A. G., Hack, A., Minekus, M., Zeijdner, E., Schoeters, G., Verstraete, W., Wiele, T. V. D., Wragg, J., Rompelberg, C. J. M., Sips, A. J. A. M. and Wijnen, J. H. V., *Environ. Sci. Technol.*, **36**, 3326–3334 (2002).
24. Rotard, W., Christmann, W., Knoth, W. and Mailahn, W., *UWSF-Z. Umweltchem. Okotox.*, **7**, 3–9 (1995).
25. Minekus, M., Marteau, P., Havenaar, R. and Huisin't Veld, J. H. J., *ALTA*, **23**, 197–216 (1995).
26. Hack, A. and Selenka, F., *Toxicol. Lett.*, **88**, 199–210 (1996).
27. Oomen, A. G., Sips, A. J. A. M., Groten, J. P., Sijm, D. T. H. M. and Tolls, J., *Environ. Sci. Technol.*, **34**, 297–303 (2000).
28. Oomen, A. G., Tolls, J., Kruidenier, M., Bosgra, S. S. D., Sips, A. J. A. M. and Groten, J. P., *Environ. Health Perspect.*, **109**, 731–737 (2001).
29. Oomen, A. G., Rompelberg, C. J. M., Van de Kamp, E., Pereboom, D. P. K. H., De Zwart, L. L. and Sips, A. J. A. M., *Arch. Environ. Contam. Toxicol.*, **46**, 183–188 (2004).
30. Holman, H. N., Goth-Goldstein, R., Aston, D., Yun, M. and Kengsoontra, J., *Environ. Sci. Technol.*, **36**, 1281–1286 (2002).
31. Ruby, M. V., Fehling, K. A., Paustenbach, D. J., Landenberger, B. D. and Holsapple, M. P., *Environ. Sci. Technol.*, **36**, 4905–4911 (2002).
32. Van de Wiele, T. R., Verstraete, W. and Siciliano, S. D., *J. Environ. Qual.*, **33**, 1343–1353 (2004).
33. Pu, X., Lee, L. S., Galinsky, R. E. and Carlson, G. P., *Toxicol. Sci.*, **79**, 10–17 (2004).
34. Pu, X., Lee, L. S., Galinsky, R. E. and Carlson, G. P., *Toxicology*, **217**, 14–21 (2006).
35. Tang, X., Tang, L., Zhu, Y., Xing, B., Duan, J. and Zheng, M., *Environ. Pollut.*, **140**, 279–285 (2006).
36. Scott, W. C. and Dean, J. R., *J. Environ. Monit.*, **7**, 710–714 (2005).
37. Esteve-Turrillas, F. A., Scott, W. C., Pastor, A. and Dean, J. R., *J. Environ. Monit.*, **7**, 1093–1098 (2005).

Chapter 7

Selected Case Studies on Bioavailability, Bioaccessibility and Mobility of Environmental Contaminants

This chapter contains details of several case studies that seek to demonstrate the bioavailability and oral bioaccessibility of metals and persistent organic pollutants from plants and soil samples. All examples have been acquired from the author's own laboratory. This does not mean that the scientific literature does not contain suitable examples – far from it. However, for ease of use to demonstrate the range of activities that this book has sought to encourage it was simpler to select specific examples with which the author was familiar.

7.1 Bioavailability of Metals by Plants[†]

7.1.1 Background

This case study investigates the uptake by lettuce, spinach, radish and carrot of Cd, Cu, Mn, Pb and Zn from compost under greenhouse conditions. The plants were harvested at maturity and their roots and leaves collected. Samples (soil and plant material) were acid-digested and analysed to determine total metal concentration. Metal concentrations in soil and plant samples (Mn and Zn) were determined by using flame atomic absorption spectroscopy (FAAS), while

[†] From Intawongse, M. and Dean, J. R., *Food Addit. Contam.*, **23**, 36–48 (2006).

differential-pulse anodic stripping voltammetry (DP-ASV) was used to determine the concentration of Cd, Cu and Pb in plant materials.

7.1.2 Experimental

7.1.2.1 Reagents

All chemicals used were of analytical grade. Copper oxide, lead nitrate and zinc oxide were provided by BDH Chemicals Ltd (Poole, Dorset, UK). Manganese metal was provided by Merck (Poole, Dorset, UK) and cadmium nitrate by Fisher Scientific UK Ltd (Loughborough, Leicestershire, UK). Concentrated hydrochloric acid and 30% hydrogen peroxide were provided by BDH Chemicals Ltd (Poole, Dorset, UK) and concentrated nitric acid by Fisher Scientific UK Ltd (Loughborough, Leicestershire, UK).

7.1.2.2 Preparation of Heavy-Metal-Contaminated Soils

Before use, the compost soil (Levington multipurpose compost) was sieved (2 mm) and air-dried (48 h). Soils were 'spiked' at two concentration levels (low and high). A control soil (unadulterated) was also used in the study. For the low-concentration level (ten times the unadulterated concentration), metal solutions at the approximate levels (10, 150, 300, 10 and 100 $\mu g\,g^{-1}$ for Cd, Cu, Mn, Pb and Zn, respectively) were prepared to 'spike' into the soil. It is noted that Cd was 'spiked' in the compost used for growing radish and carrot only. The soil was weighed, approximately 300 g for each mixing, and placed in a stainless-steel tray. Then, the soil was thoroughly mixed with the metal solution to ensure homogeneity and air-dried for a day to allow excessive water to evaporate. The high-concentration level (fifty times the unadulterated concentration) was prepared in the same manner as described above.

7.1.2.3 General Procedure: Growing Vegetable Plants

Seeds of spinach, lettuce, radish and carrot (obtained directly from local markets) were germinated in plastic trays and the seedlings transplanted after two weeks, into individual plastic pots containing 100 g of metal-contaminated soil. The plants were grown in soil contaminated at low and high concentrations for both individual and mixed metals with five pots per treatment. Three plants of spinach, lettuce, radish and carrot were also planted in unadulterated soil as control samples. The plants were frequently watered with distilled water; the unadulterated and 'low-treatment' plants were watered daily and the 'high-treatment' plants two to three times a week. The plants were grown under artificial light (sodium lighting system, 150 W m^{-2}) with time-intervals of 16 h daylight and 8 h dark. The temperature was within the range 12.1 to 25.3°C, while the humidity varied between 38 and 89%. The mature plants (6–8 weeks growth) were harvested. Roots and leaves were separated and then thoroughly washed, including a final rinse with distilled water, and kept in a freezer at $-18°C$ until analysis. Soil samples beneath each plant root were collected after harvesting the plant. The Cd, Cu, Mn, Pb and Zn determinations were carried out on plant (roots and leaves) and soil samples.

7.1.2.4 Soil-Digestion Procedure

For determination of total heavy-metal content of the soils, 1 g of each soil sample (oven-dried at 70°C for 48 h) was accurately weighed into a digestion tube and 10 ml of concentrated nitric acid:water (1:1 (vol/vol)) added. The sample was then heated at 95°C on a heating block (2006 Digester, Foss Tecator, Hoganas, Sweden) for 15 min without boiling. After cooling at room temperature for 5 min, 5 ml of concentrated HNO_3 was added and the sample were heated at 95°C for 30 min. Additional 5 ml aliquots of concentrated HNO_3 were added until no brown fumes were given off. The solution was allowed to evaporate to < 5 ml. After cooling, 2 ml of water and 3 ml of 30% H_2O_2 were added and heated (<bpt) until effervescence subsided and the solution cooled. Additional H_2O_2 was added until effervescence ceased (but no more than 10 ml H_2O_2 was added). This stage was continued for 2 h at a temperature less than the boiling point. Then, the solution was allowed to evaporate to < 5 ml. After cooling, 10 ml of concentrated HCl were added and the solution was heated at 95°C for 15 min. After cooling, the sample was filtered through a Whatman No. 41 filter paper into a 100 ml volumetric flask and then 'made up to the mark' with distilled water.

7.1.2.5 Plant-Digestion Procedure

Into a digestion tube, 1 g (accurately weighed) of plant sample (oven-dried at 70°C for 48 h) was placed and 10 ml of concentrated HNO_3 added. The sample was then heated to 95°C on the heating block for ca. 1 h. After cooling, 5 ml of concentrated H_2SO_4 were added and the sample was heated to 140°C until charring first occurred. After cooling, 5 ml of concentrated HNO_3 were added and heated to 180°C. Further aliquots of HNO_3 were added until the sample digest appeared clear or a pale-straw colour. After cooling, 1 ml of $500\,g\,l^{-1}$ H_2O_2 was added and heated to 200°C. This procedure was repeated until brown fumes ceased to appear. After cooling, 10 ml of distilled water and 0.5 ml of concentrated HNO_3 were added and heated to 200°C until white fumes were evolved. After cooling, 10 ml of distilled water and 1 ml of $500\,g\,l^{-1}$ H_2O_2 were added and heated to 240°C until white fumes were evolved. Finally, the digest was cooled and filtered through a Whatman No. 41 filter paper into a 50 ml volumetric flask and then 'made up to the mark' with distilled water.

7.1.2.6 FAAS Analysis

Soil (Cd, Cu, Mn, Pb and Zn) and plant (Mn and Zn) samples were analysed by using an atomic absorption spectrometer (AAnalyst 100, Perkin-Elmer, Norwalk, CT, USA) with an air–acetylene flame. The measurements were made under operating conditions for the most sensitive absorption lines of individual elements, i.e. Cd, Cu, Mn, Pb and Zn at 228.8, 324.8, 279.5, 217.0 and 213.9 nm, respectively, with a hollow-cathode-lamp current at 7 mA and a slit width of 0.7 nm.

7.1.2.7 DP-ASV Analysis

A 757 VA Computrace (Metrohm, Herisau, Switzerland) voltameter in the differential-pulse–anodic stripping voltammetry (DP-ASV) mode was used to determine Cd, Cu and Pb in plant samples. The voltameter was equipped with a three-electrode system: a hanging mercury-dropping electrode (HMDE) as the working electrode, a Pt electrode as the auxiliary electrode and a KCl-saturated Ag/AgCl electrode as the reference electrode.

All electrodes and a measuring cell were thoroughly cleaned with distilled water. The measurements were carried out by pipetting a 1 ml portion of the plant extract into a measuring vessel containing 9 ml of distilled water. Then, 1 ml of the electrolyte solution (NaOAc) was added. The electrode was immersed in the solution and then scanned from -1.0 to 0.0 V. Nitrogen gas was used as the purging gas at 1 bar pressure and the measurements were carried out in the 'differential-pulse' (DP) mode. The standard-additions method was applied by analysing five calibration standard solutions of a 1 ppm stock solution of each metal. A blank was analysed with each analytical batch.

7.1.3 Results and Discussion

The mean concentrations of metals in the roots and leaves of four vegetables grown on different treatments of contaminated soils are shown in Table 7.1. Lettuce and spinach represented the edible vegetable leaves (above ground) and radish and carrot represented edible vegetable roots (underground). Control plants were also produced by growing on uncontaminated soil. It was observed that each metal differed considerably in uptake from each other.

7.1.3.1 Cadmium

It should be noted that the Cd treatments were undertaken for only radish and carrot. From the results obtained, the cadmium content in leaves of both radish and carrot were significantly higher than in their roots, except for the unadulterated treatment which was not significantly different. The highest Cd contents were found in Cd–H treatments for radish leaves, radish roots and carrot leaves, with concentrations of 184.4, 68.2 and 47.4 μg g^{-1}, respectively. The lowest Cd concentration was in carrot leaves $(0.2\,\mu$g g$^{-1})$ in the unadulterated soil.

7.1.3.2 Copper

Copper uptake was significantly higher in the roots than in the leaves for lettuce and spinach from every treatment, i.e. unadulterated, Mix–L, Mix–H, Zn–L, Zn–H, Cu–L and Cu–H. In contrast, copper concentrations in radish and carrot

Table 7.1 Metal concentrations (μg g⁻¹) of compost used for growing (a) lettuce, (b) spinach, (c) radish and (d) carrot under different treatments[a,b,c]

Element	(a) Mean concentration ($\mu g\,g^{-1}$, dry weight): soil batch #1, used for growing lettuce						
	Control (n = 3)	Mix–L (n = 5)	Mix–H (n = 3)[d]	Zn–L (n = 5)	Zn–H (n = 3)	Cu–L (n = 5)	Cu–H (n = 3)
Cu	11.3 ± 1.3	279.4 ± 4.4	—	13.4 ± 4.8	10.6 ± 1.4	207.9 ± 11.0	927.9 ± 24.9
Mn	45.3 ± 2.5	572.8 ± 23.2	—	48.4 ± 4.3	41.5 ± 7.5	51.6 ± 3.3	36.9 ± 5.5
Pb	6.6 ± 4.6	29.5 ± 2.2	—	8.8 ± 1.8	8.0 ± 8.0	8.0 ± 6.3	ND
Zn	17.7 ± 5.0	176.8 ± 4.0	—	117.0 ± 5.8	522.3 ± 10.4	15.7 ± 2.6	14.6 ± 0.4

Element	(b) Mean concentration ($\mu g\,g^{-1}$, dry weight): soil batch #2, used for growing spinach						
	Control (n = 3)	Mix–L (n = 5)	Mix–H (n = 3)	Zn–L (n = 5)	Zn–H (n = 3)	Cu–L (n = 5)	Cu–H (n = 3)
Cu	17.4 ± 1.2	336.3 ± 7.2	1205.0 ± 57.8	16.1 ± 2.9	19.8 ± 0.0	239.1 ± 14.7	1626.1 ± 72.5
Mn	51.5 ± 2.1	508.3 ± 73.0	1939.4 ± 60.0	49.4 ± 2.6	56.1 ± 0.1	55.6 ± 4.6	44.9 ± 0.0
Pb	11.9 ± 4.0	33.2 ± 2.2	92.4 ± 9.0	14.2 ± 3.5	15.8 ± 7.9	12.6 ± 1.8	13.2 ± 4.6
Zn	13.8 ± 2.1	181.7 ± 4.7	581.8 ± 2.8	112.6 ± 6.0	531.9 ± 5.5	15.0 ± 1.6	13.4 ± 1.3

(continued overleaf)

Table 7.1 (*continued*)

(c) Mean concentration (μg g^{-1}, dry weight): soil batch #3, used for growing radish

Element	Control ($n = 3$)	Mix–L ($n = 5$)	Mix–H ($n = 4$)	Cd–L ($n = 4$)	Cd–H ($n = 4$)
Cd	0.4 ± 0.3	10.4 ± 1.0	49.1 ± 1.0	11.0 ± 0.3	70.6 ± 3.2
Cu	16.7 ± 2.1	170.7 ± 3.3	793.4 ± 31.2	14.6 ± 0.0	16.7 ± 0.0
Mn	59.2 ± 0.1	381.9 ± 6.5	1829.3 ± 13.7	62.2 ± 4.2	60.0 ± 1.4
Pb	12.5 ± 0.0	22.6 ± 3.5	58.0 ± 3.1	14.1 ± 3.1	12.6 ± 5.1
Zn	8.7 ± 1.1	93.5 ± 1.8	490.3 ± 5.4	10.0 ± 1.0	13.1 ± 0.7

(d) Mean concentration (μg g^{-1}, dry weight): soil batch #4, used for growing carrot

Element	Control ($n = 1$)	Mix–L ($n = 3$)	Cd–L ($n = 2$)	Cd–H ($n = 3$)
Cd	0.0	10.8 ± 0.7	7.1 ± 1.7	39.6 ± 2.4
Cu	20.9	180.5 ± 4.3	18.8 ± 0.0	19.5 ± 1.2
Mn	65.0	388.3 ± 20.6	60.7 ± 2.1	68.1 ± 15.6
Pb	6.3	31.4 ± 6.3	25.1 ± 0.0	16.7 ± 3.6
Zn	13.3	111.8 ± 0.7	9.5 ± 1.6	10.8 ± 0.6

[a] Control, no 'spiking'; L, 'low-treatment'; H, 'high-treatment'; Mix–L, mixed metal at 'low-level' treatment;
Mix–H, mixed metal at 'high-level' treatment.
[b] ND, not detectable.
[c] From Intawongse, M. and Dean, J. R., *Food Adit. Contam.*, **23**, 36–48 (2006).
[d] No plants survived and soils were not collected.

leaves from every treatment (except the leaves of radish in the Cd–H treatment) were higher than in the roots. Copper concentrations were highest in lettuce roots ($262.1\,\mu g\,g^{-1}$) in the Cu–H treatment, followed by spinach roots in the Cu–H treatment ($150.1\,\mu g\,g^{-1}$) and the Mix–H treatment ($70.1\,\mu g\,g^{-1}$), and lettuce roots in the Mix–L treatment ($69.0\,\mu g\,g^{-1}$). The Cu content was lowest in radish roots ($2.1\,\mu g\,g^{-1}$) in the Cd–L treatment.

7.1.3.3 Manganese

Manganese is the most available metal. The concentration of Mn in the plants ranged from 16.9 to $6631.6\,\mu g\,g^{-1}$. Plant leaves contain more manganese than their roots, except spinach leaves in the unadulterated and Cu–L treatment. Manganese concentrations were highest in spinach leaves ($6631.6\,\mu g\,g^{-1}$) in the Mix–H treatment, followed by radish leaves in the Mix–H treatment ($3821.8\,\mu g\,g^{-1}$) and spinach leaves in the Mix–L treatment ($1971.1\,\mu g\,g^{-1}$). The Mn content was lowest in radish roots ($16.9\,\mu g\,g^{-1}$) in the Cd–H treatment.

7.1.3.4 Lead

Lead is the least available metal considered. Lead concentrations varied from the non-detectable level to $11.8\,\mu g\,g^{-1}$ and significantly lower than other metals accumulated in the plants. This can be explained by the fact that lead binds to organic matter in soil, limiting uptake by the plants [1]. Lead contents in every treatment, except in radish roots in the Cd–H treatment ($11.8\,\mu g\,g^{-1}$) and spinach roots in the Mix–H treatment ($11.1\,\mu g\,g^{-1}$), were lower than $10\,\mu g\,g^{-1}$.

7.1.3.5 Zinc

Zinc concentrations in leaves of radish, carrot and spinach were significantly higher than in their roots for every treatment, except the spinach leaves in the Zn–H and Cu–H treatments, which were slightly lower. In contrast, the zinc contents in lettuce leaves were lower than in their roots for every treatment, except in the case of the Cu–H treatment which was not significantly different in concentrations between their leaves and roots. The highest zinc contents were found in the Zn–H treatments for lettuce roots, lettuce leaves, spinach roots and spinach leaves, with concentrations of 1440.6, 1171, 719 and $632\,\mu g\,g^{-1}$, respectively. This is in agreement with a previous study showing that lettuce and spinach accumulated zinc to a greater extent [2]. The Zn content was lowest in carrot roots ($23.6\,\mu g\,g^{-1}$) from the Cd–L treatment.

7.1.3.6 Influence of Plant Species

Table 7.2 indicates that individual plant types differ in their metal uptake. The Cu uptake was slightly low in lettuce and carrot, but relatively higher in spinach and radish. Manganese uptake was significantly higher in lettuce and spinach and slightly lower in radish and carrot. Zinc uptake was relatively high, whereas Pb tended to be taken up least from all of the plants studies. Cadmium uptake was rather high in radish, but lower in carrot.

Table 7.2 Metal concentrations (μg g^{-1}, dry plant material) found in (a) lettuce, (b) spinach, (c) radish, and (d) carrot[a,b,c]

(a) Mean concentration (μg g^{-1}): lettuce grown on soil batch #1

Element	Control Leaves (n = 3)	Control Roots (n = 3)	Mix-L Leaves (n = 4)	Mix-L Roots (n = 4)	Mix-H[d] Leaves	Mix-H[d] Roots	Zn-L Leaves (n = 4)	Zn-L Roots (n = 4)	Zn-H Leaves (n = 3)	Zn-H Roots (n = 1)	Cu-L Leaves (n = 4)	Cu-L Roots (n = 4)	Cu-H Leaves (n = 3)	Cu-H Roots (n = 3)
Cu	2.4 ± 0.2	10.4 ± 7.7	13.5 ± 2.2	69.0 ± 25.5	—	—	3.6 ± 1.5	25.4 ± 13.8	8.0 ± 0.7	21.1	7.1 ± 1.6	25.2 ± 4.2	12.8 ± 2.6	262.1 ± 36.9
Mn	114.3 ± 9.8	65.2 ± 12.3	1246.6 ± 161.9	1117.6 ± 461.0	—	—	148.9 ± 33.8	74.3 ± 9.3	900.1 ± 314.2	221.3	108.5 ± 24.7	88.9 ± 51.3	381.3 ± 71.4	331.3 ± 19.9
Pb	0.2 ± 0.2	1.0 ± 1.2	ND	ND	—	—	0.3 ± 0.1	1.1 ± 0.4	1.7 ± 3.0	ND	2.1 ± 2.8	1.0 ± 0.7	4.3 ± 5.3	0.6 ± 1.0
Zn	31.3 ± 8.1	42.6 ± 17.1	154.3 ± 43.8	194.0 ± 56.0	—	—	78.3 ± 22.9	97.5 ± 16.7	1171.0 ± 236.9	1440.6	30.2 ± 3.2	46.7 ± 15.5	54.7 ± 4.2	51.6 ± 5.3

(b) Mean concentration (μg g^{-1}): spinach grown on soil batch #2

Element	Control Leaves (n = 3)	Control Roots (n = 3)	Mix-L Leaves (n = 5)	Mix-L Roots (n = 5)	Mix-H Leaves (n = 1)	Mix-H Roots (n = 1)	Zn-L Leaves (n = 4)	Zn-L Roots (n = 4)	Zn-H Leaves (n = 3)	Zn-H Roots (n = 3)	Cu-L Leaves (n = 4)	Cu-L Roots (n = 4)	Cu-H Leaves (n = 3)	Cu-H Roots (n = 3)
Cu	4.6 ± 0.3	21.2 ± 5.1	20.5 ± 3.5	46.1 ± 5.4	29.6	70.1	6.0 ± 2.4	11.7 ± 4.7	7.5 ± 3.0	43.0 ± 18.9	17.9 ± 4.3	31.3 ± 15.0	32.3 ± 5.9	150.1 ± 50.3
Mn	180.7 ± 18.5	232.5 ± 44.9	1971.1 ± 324.4	882.7 ± 291.3	6631.6	901.5	246.6 ± 49.1	221.0 ± 43.4	233.3 ± 51.2	229.2 ± 71.4	206.0 ± 32.5	226.2 ± 64.1	468.4 ± 103.4	260.4 ± 72.7
Pb	0.2 ± 0.1	0.7 ± 0.4	1.7 ± 2.8	4.0 ± 3.8	5.2	11.1	1.9 ± 1.3	0.3 ± 0.5	0.3 ± 0.6	3.9 ± 4.3	3.6 ± 1.0	7.4 ± 6.8	1.0 ± 0.9	7.8 ± 0.5
Zn	70.2 ± 12.9	67.2 ± 5.8	233.5 ± 60.3	162.2 ± 36.5	638.5	275.9	225.9 ± 69.3	167.0 ± 32.0	632.0 ± 29.9	719.0 ± 175.6	45.6 ± 4.3	42.2 ± 4.6	54.9 ± 3.2	62.4 ± 9.8

Table 7.2 (*continued*)

(c) Mean concentration ($\mu g\,g^{-1}$): radish grown on soil batch #3

Element	Control Leaves (n = 3)	Control Roots (n = 3)	Mix–L Leaves (n = 5)	Mix–L Roots (n = 5)	Mix–H Leaves (n = 1)	Mix–H Roots (n = 1)	Cd–L Leaves (n = 3)	Cd–L Roots (n = 3)	Cd–H Leaves (n = 4)	Cd–H Roots (n = 4)
Cd	0.4 ± 0.4	0.9 ± 0.4	9.8 ± 1.8	5.7 ± 1.2	44.8	19.2	29.5 ± 12.8	8.8 ± 2.2	184.4 ± 6.4	68.2 ± 68.0
Cu	15.0 ± 7.1	10.7 ± 3.0	19.8 ± 4.3	7.2 ± 1.9	48.6	26.9	11.9 ± 9.1	2.1 ± 0.9	6.0 ± 3.6	36.3 ± 30.8
Mn	104.1 ± 37.0	17.6 ± 3.5	510.3 ± 114.8	60.5 ± 8.6	3821.8	758.9	139.4 ± 16.3	17.2 ± 6.7	134.7 ± 25.7	16.9 ± 19.8
Pb	3.4 ± 0.6	4.4 ± 4.2	0.8 ± 1.9	ND	ND	ND	2.7 ± 1.8	ND	0.9 ± 0.1	11.8 ± 13.1
Zn	32.2 ± 8.2	25.6 ± 5.9	121.7 ± 19.2	69.1 ± 7.1	599.3	500.3	26.9 ± 8.5	25.9 ± 6.7	62.6 ± 4.8	36.7 ± 37.1

(d) Mean concentration ($\mu g\,g^{-1}$): carrot grown on soil batch #4

Element	Control Leaves (n = 1)	Control Roots (n = 1)	Mix–L Leaves (n = 3)	Mix–L Roots (n = 1)	Cd–L Leaves (n = 2)	Cd–L Roots (n = 2)	Cd–H Leaves (n = 3)	Cd–H Roots (n = 3)
Cd	0.2	1.0	8.3 ± 4.6	9.5	10.5 ± 3.2	8.0 ± 1.8	47.4 ± 6.8	27.4 ± 5.2
Cu	12.5	6.4	24.9 ± 12.7	15.5	10.1 ± 2.1	5.8 ± 0.9	16.3 ± 8.0	8.3 ± 7.7
Mn	198.0	34.4	1857.1 ± 716.6	271.0	165.5 ± 31.4	27.7 ± 3.0	342.6 ± 101.1	56.3 ± 19.7
Pb	0.7	ND	8.2 ± 11.3	ND	ND	0.5 ± 0.7	7.3 ± 8.8	1.2 ± 0.8
Zn	41.8	25.4	96.9 ± 10.8	52.6	40.0 ± 2.5	23.6 ± 4.5	43.2 ± 8.2	26.7 ± 5.0

[a] Control, no 'spiking'; L, 'low-treatment'; H, 'high-treatment'; Mix–L, mixed metal at 'low-level' treatment; Mix–H, mixed metal at 'high-level' treatment.
[b] ND, not detectable.
[c] From Intawongse, M. and Dean, J. R., *Food Addit. Contam.*, **23**, 36–48 (2006).
[d] No plants survived.

7.1.3.7 Transfer Factors (TFs) of Heavy Metals from Soil to Plants

One approach to assess the mobilities of metals by plants is to calculate the transfer factor (TF), as defined in the following equation:

$$TF = \frac{C_{plant}}{C_{total-soil}} \tag{7.1}$$

where C_{plant} is the concentration of an element in the plant material (dry-weight basis) and $C_{total-soil}$ is the total concentration of the same element in the soil (dry-weight basis) where the plant was grown. The higher the value of the TF, then the more mobile/available the metal is. The TF values of the metals for the plants studied are presented in Table 7.3. The results indicate that the TF values for Cd, Cu, Mn, Pb and Zn for various vegetables varied greatly between plant types and soil treatments. The TF values for Cd were generally slightly higher than those for Cu, but lower than Zn. The TF values for Cu varied from 0.21 (lettuce leaves) to 1.22 (spinach roots) in the unadulterated soil and 0.01 (lettuce leaves, Cu–H treatment) to 2.17 (spinach roots, Zn–H treatment and radish roots, Cd–H treatment) in contaminated soil. The values for Pb were much lower than those for Cu, varying from 0.00 (carrot roots) to 0.35 (radish roots) in unadulterated soil and 0.00 (leaves and roots of lettuce, radish and carrot) to 0.94 (radish roots, Cd–H treatment). The TF values for Mn varied the most, with values from 0.30 (radish roots) to 4.52 (spinach roots) in unadulterated soil and 0.16 (radish roots, Mix–L treatment) to 21.66 (lettuce leaves, Zn–H treatment). On the whole, the order of the transfer factors was Mn $\gg$ Zn > Cd > Cu > Pb. Typical metal transfer factors are 0.0–2.7, 0.01–2.17, 0.3–21.7, 0.0–0.9 and 0.48–5.07 for Cd, Cu, Mn, Pb and Zn, respectively.

7.1.4 Conclusions

The results showed that the uptake of Cd, Cu, Mn and Zn by plants corresponded to the increasing level of soil contamination, while the uptake of Pb was low. Soil-to-plant transfer factor (TF) values decreased from Mn $\gg$ Zn > Cd > Cu > Pb. Moreover, it was observed from this investigation that individual plant types greatly differ in their metal uptake, e.g. spinach accumulated a high content of Mn and Zn, while relatively lower concentrations were found for Cu and Pb in their tissues.

7.1.5 Specific References

1. Wang, X.-P., Shan, X.-Q., Zhang, S.-Z. and Wen, B., *Chemosphere*, **55**, 811–822 (2004).

2. Sauerbeck, D. R., *Water, Air Soil Pollut.*, **57–58**, 227–237 (1991).

Table 7.3 Ratio of concentrations of metals in plants to metals in soil: transfer factor (TF) values for (a) lettuce, (b) spinach, (c) radish and (d) carrot[a,b]

(a) TF value: lettuce grown on soil batch #1

Element	Control (n = 3)		Mix–L (n = 4)		Mix–H[c]		Zn–L (n = 4)		Zn–H (n = 3)		Cu–L (n = 4)		Cu–H (n = 3)	
	Leaves	Roots	Leaves	Roots	Leaves	Roots	Leaves	Roots	Leaves	Roots	Leaves	Roots	Leaves	Roots
Cu	0.21	0.92	0.05	0.25	—	—	0.27	1.89	0.76	2.00	0.03	0.12	0.01	0.28
Mn	2.52	1.44	2.18	1.95	—	—	3.08	1.53	21.66	5.33	2.10	1.72	10.35	8.99
Pb	0.03	0.15	0.00	0.00	—	—	0.04	0.13	0.66	0.00	0.27	0.12	0.00	0.00
Zn	1.78	2.41	0.87	1.10	—	—	0.67	0.83	2.24	2.76	1.93	2.98	3.75	3.54

(b) TF value: spinach grown on soil batch #2

Element	Control (n = 3)		Mix–L (n = 5)		Mix–H (n = 1)		Zn–L (n = 4)		Zn–H (n = 3)		Cu–L (n = 4)		Cu–H (n = 3)	
	Leaves	Roots	Leaves	Roots	Leaves	Roots	Leaves	Roots	Leaves	Roots	Leaves	Roots	Leaves	Roots
Cu	0.26	1.22	0.06	0.14	0.02	0.06	0.37	0.72	0.38	2.17	0.07	0.13	0.02	0.09
Mn	3.51	4.52	3.88	1.74	3.42	0.46	4.99	4.47	4.16	4.09	3.71	4.07	10.44	5.80
Pb	0.02	0.06	0.05	0.12	0.06	0.12	0.13	0.02	0.02	0.25	0.28	0.59	0.08	0.60
Zn	5.07	4.86	1.29	0.89	1.10	0.47	2.01	1.48	1.19	1.35	3.03	2.81	4.10	4.66

(continued overleaf)

Table 7.3 (*continued*)

(c) TF value: radish grown on soil batch #3

Element	Control ($n = 3$)		Mix–L ($n = 5$)		Mix–H ($n = 1$)		Cd–L ($n = 3$)		Cd–H ($n = 4$)	
	Leaves	Roots	Leaves	Roots	Leaves	Roots	Leaves	Roots	Leaves	Roots
Cd	0.91	2.25	0.95	0.55	0.91	0.39	2.68	0.80	2.61	0.97
Cu	0.90	0.64	0.12	0.04	0.06	0.03	0.81	0.15	0.36	2.17
Mn	1.76	0.30	1.34	0.16	2.09	0.41	2.24	0.28	2.25	0.28
Pb	0.27	0.35	0.19	0.00	0.00	0.00	0.19	0.01	0.07	0.94
Zn	3.69	2.93	1.30	0.74	1.22	1.02	2.68	2.58	4.78	2.81

(d) TF value: carrot grown on soil batch #4

Element	Control ($n = 1$)		Mix–L ($n = 3$)		Cd–L ($n = 2$)		Cd–H ($n = 3$)	
	Leaves	Roots	Leaves	Roots	Leaves	Roots	Leaves	Roots
Cd	0.00	0.00	0.77	0.88	1.48	1.14	1.20	0.69
Cu	0.60	0.31	0.14	0.09	0.54	0.31	0.83	0.43
Mn	3.04	0.53	4.78	0.70	2.73	0.46	5.03	0.83
Pb	0.11	0.00	0.26	0.00	0.00	0.04	0.44	0.07
Zn	3.13	1.90	0.87	0.47	4.23	2.49	4.01	2.48

[a]Control, no 'spiking'; L, 'low-treatment'; H, 'high-treatment'; Mix–L, mixed metal at 'low-level' treatment; Mix–H, mixed metal at 'high-level' treatment.
[b]From Intawongse, M. and Dean, J. R., *Food Addit. Contam.*, **23**, 36–48 (2006).
[c]No plants survived.

7.2 Bioaccessibility of Metals from Plants[‡]

7.2.1 Background

This case study investigates the oral bioaccessibility of selected metals (Cd, Cu, Mn, Pb and Zn) from plants (lettuce, spinach, radish and carrot) which had been previously grown on contaminated compost. The edible part of plants, i.e. the leaves of lettuce and spinach and the roots of radish and carrot, were extracted using an *in vitro* gastrointestinal (GI) extraction to assess metal bioavailability. Flame atomic absorption spectroscopy (FAAS) was employed to determine metal concentrations in soil and plant samples (Mn and Zn), while Cd, Cu and Pb in plant materials were analysed by differential pulse–anodic stripping voltammetry (DP-ASV).

7.2.2 Experimental

7.2.2.1 Reagents

All chemicals used were of analytical grade. Copper oxide, lead nitrate and zinc oxide were provided by BDH Chemicals Ltd (Poole, Dorset, UK). Manganese metal was provided by Merck (Poole, Dorset, UK) and cadmium nitrate by Fisher Scientific UK Ltd (Loughborough, Leicestershire, UK). Concentrated hydrochloric acid and 30% hydrogen peroxide were provided by BDH Chemicals Ltd (Poole, Dorset, UK) and concentrated nitric acid by Fisher Scientific UK Ltd (Loughborough, Leicestershire, UK). Pepsin-A powder (1 anson unit per g) (lactose as diluent), amylase and pancreatin were provided by BDH Chemicals Ltd (Poole, Dorset, UK) and bile salts by Sigma-Aldrich Company Ltd (Gillingham, Dorset, UK).

The tea leaf Certified Reference Material (CRM) (INCT-TL-1) was obtained from the Institute of Nuclear Chemistry and Technology, Warsaw, Poland.

7.2.2.2 In vitro *Gastrointestinal Extraction of Vegetable Plants*

This determination consists of two sequential processes, namely a gastric and an intestinal digestion, with each one carried out employing simulated human conditions (enzymes, pH and temperature). In the first stage, approximately 1 g (accurately weighed) of plant samples (oven-dried at 70°C for 48 h) was placed into a 50 ml Sarstedt tube and treated with 15 ml of pepsin (1% (wt/vol)) in saline (154 mmol l^{-1}). The pH of the solution was adjusted to pH 1.8 with dilute HCl. The mixture was then shaken at 100 rpm in a thermostatic bath maintained at 37°C. After 4 h, the solution was centrifuged at 3000 rpm for 10 min and the supernatant was removed and its pH adjusted to 2.5 with 2 M HCl.

The second stage involved extraction with intestinal juices. To the gastric digest residue, 7.5 ml of a mixed solution of pancreatin (3% (wt/vol)) and amylase (1%

[‡] From Intawongse, M. and Dean, J. R., *Food Addit. Contam.*, **23**, 36–48 (2006).

(wt/vol)) in saline and 7.5 ml of bile salts (0.15% (wt/vol)) in saline were added. The sample pH was adjusted to 7.0 with saturated $NaHCO_3$. The sample was then shaken at 100 rpm in a thermostatic bath maintained at 37°C. After 4 h, the solution was centrifuged at 3000 rpm for 10 min and the supernatant was removed. All extracts (gastric and intestinal) were analysed by either DP-ASV or FAAS. The resultant sample residue was further extracted by acid-digestion and analysed by DP-ASV for Cd, Cu and Pb and FAAS for Mn and Zn.

7.2.2.3 Plant-Digestion Procedure

Into a digestion tube, 1 g (accurately weighed) of plant sample (oven-dried at 70°C for 48 h) was placed and 10 ml of concentrated HNO_3 added. The sample was then heated to 95°C on the heating block for ca. 1 h. After cooling, 5 ml of concentrated H_2SO_4 were added and the sample was heated to 140°C until charring first occurred. After cooling, 5 ml of concentrated HNO_3 were added and heated to 180°C. Further aliquots of HNO_3 were added until the sample digest appeared clear or a pale-straw colour. After cooling, 1 ml of $500\,g\,l^{-1}$ H_2O_2 was added and heated to 200°C. This procedure was repeated until brown fumes ceased to appear. After cooling, 10 ml of distilled water and 0.5 ml of concentrated HNO_3 were added and heated to 200°C until white fumes were evolved. After cooling, 10 ml of distilled water and 1 ml of $500\,g\,l^{-1}$ H_2O_2 were added and heated to 240°C until white fumes were evolved. Finally, the digest was cooled and filtered through a Whatman No 41 filter paper into a 50 ml volumetric flask and then 'made up to the mark' with distilled water.

7.2.2.4 FAAS Analysis

Plant (Mn and Zn) samples were analysed by using an atomic absorption spectrometer (AAnalyst 100, Perkin-Elmer, Norwalk, CT, USA) with an air–acetylene flame. The measurements were made under operating conditions for the most sensitive absorption lines of individual elements, i.e. Mn and Zn at 279.5 and 213.9 nm, respectively, with a hollow-cathode-lamp current at 7 mA and a slit width of 0.7 nm.

7.2.2.5 DP-ASV Analysis

A 757 VA Computrace (Metrohm, Herisau, Switzerland) voltameter in the differential-pulse–anodic stripping voltammetry (DP-ASV) mode was used to determine Cd, Cu and Pb in plant samples. The voltameter was equipped with a three-electrode system: a hanging mercury-dropping electrode (HMDE) as the working electrode, a Pt electrode as the auxiliary electrode, and a KCl saturated Ag/AgCl electrode as the reference electrode.

All electrodes and a measuring cell were thoroughly cleaned with distilled water. The measurements were carried out by pipetting a 1 ml portion of the plant extract into a measuring vessel containing 9 ml of distilled water. Then,

1 ml of the electrolyte solution (NaOAc) was added. The electrode was immersed in the solution and then scanned from -1.0 to $0.0\,\text{V}$. Nitrogen gas was used as the purging gas at 1 bar pressure and the measurements were carried out in the 'differential-pulse' (DP) mode. The standard-additions method was applied by analysing five calibration standard solutions of a 1 ppm stock solution of each metal. A blank was analysed with each analytical batch.

7.2.3 Results and Discussion

In order to assess the oral bioavailability of metals in vegetables, the edible plant materials were extracted by using an *in vitro* gastrointestinal extraction procedure. The experimental approach consists of two processes which simulate human digestion. The first stage is designed to mimic extraction in an 'acidic stomach' using gastric fluid (pepsin), followed by stage two which involves extraction in the neutral small intestine using intestinal fluid (pancreatin, amylase and bile salts). The sample residues left from this process were then acid-digested (residual fraction). After the extractions were complete, all extracts (gastric, intestinal and residual) were analysed by either DP-ASV or FAAS. The CRM (tea leaves) was extracted and analysed in the same manner as the sample, in order to evaluate the approach. The *in vitro* results (gastric, intestinal and residual) were compared to the results obtained from acid digestion (HNO_3 plus H_2O_2) and are given in Tables 7.4 and 7.5. It is noted that Pb was not investigated in this experiment due to its low accumulation in edible plant materials; hence, it is impossible to make meaningful measurements from this *in vitro* study.

7.2.3.1 Cadmium

Cadmium treatment for the *in vitro* gastrointestinal study was applied only to radish as no carrot samples were available. The cadmium solubility was highest (54.9%) in the gastric extraction. The cadmium concentrations were similar between the intestinal phase (24.8%) and the residual fraction (20.2%).

7.2.3.2 Copper

The solubility of copper greatly differs among different plants and extraction phases. The percentages of soluble copper in lettuce were 12.6, 44.4 and 42.9 in the gastric, intestinal and residual phases, respectively. For spinach, none of the copper was soluble in the neutral-pH (intestinal) phase, 24% was found in the acidic phase (gastric fluid), while most of the copper (76%) was left in the residual phase. About 3% of the copper in radish can be recovered during intestinal extraction, while 62.5 and 34.7% were found in the gastric and residual fractions, respectively.

7.2.3.3 Manganese

Most of the manganese in lettuce (63.7%) and radish (45.8%) was extracted in the acidic phase (gastric fluid), while it was significantly lower in spinach

Table 7.4 Total metal levels (certified values and acid-digestion) and concentrations (in $\mu g\,g^{-1}$, dry weight) obtained from the PBET experiment for tea CRM (INCT-TL-1)[a]

Element[b]	Certified value Mean ± SD	Acid-digestion, Mean ± SD	Physiologically based extraction test (PBET)						Total of phases I + II + III
			Phase I (gastric)		Phase II (intestinal)[c]		Phase III (residual)		
			Mean ± SD	%	Mean ± SD	%	Mean ± SD	%	
Cu[d]	20.4 ± 1.5	24.3 ± 0.1	1.3 ± 0.4	5.4	ND	ND	23.4 ± 8.2	94.6	24.8 ± 8.0
Mn[e]	1570 ± 110	1541 ± 21	888 ± 56	52.1	80 ± 5	4.7	735 ± 80	43.2	1703 ± 146

[a]From Intawongse, M. and Dean, J. R., *Food Adit. Contam.*, **23**, 36–48 (2006).
[b]For tea leaves ($n = 3$).
[c]ND, non-detectable.
[d]Measured by DP-ASV.
[e]Measured by FAAS.

Table 7.5 Total metal levels (acid-digestion) and concentrations (in $\mu g\,g^{-1}$, dry weight) obtained from the PBET experiment for lettuce, spinach and radish leaves[a]

Vegetable	Element[b]	Acid-digestion, mean ± SD	Physiologically based extraction test (PBET)							Total of phases I + II + III
			Phase I (gastric)		Phase II (intestinal)[c]		Phase III (residual)			
			Mean ± SD	%	Mean ± SD	%	Mean ± SD	%		
Lettuce (n = 3)	Mn	1246.6 ± 161.9	768.4 ± 26.9	63.7	125.2 ± 7.7	10.4	312.2 ± 10.8	25.9		1205.8 ± 34.9
	Zn	154.3 ± 43.8	80.3 ± 14.6	45.2	12.3 ± 5.8	6.9	84.9 ± 1.7	47.8		177.5 ± 20.1
Spinach (n = 2)	Cu	20.5 ± 3.5	4.0 ± 1.8	24.0	ND	ND	12.8 ± 3.1	76.0		16.9 ± 3.6
	Mn	1971.1 ± 324.4	347.1 ± 98.7	16.5	182.6 ± 44.1	8.7	1567.7 ± 271.0	74.7		2097.5 ± 89.3
	Zn	233.5 ± 60.3	16.7 ± 6.8	7.0	12.8 ± 15.1	5.3	211.2 ± 44.1	87.7		240.8 ± 67.5
Radish (n = 3)	Cd	5.7 ± 1.2	3.0 ± 0.4	54.9	1.3 ± 0.5	24.8	1.1 ± 0.3	20.2		5.4 ± 1.1
	Mn	60.5 ± 8.6	26.4 ± 1.5	45.8	11.9 ± 1.0	20.6	19.4 ± 3.0	33.7		57.7 ± 6.8
	Zn	69.1 ± 7.1	21.2 ± 6.3	32.3	6.0 ± 2.4	9.2	38.5 ± 2.2	58.5		65.7 ± 10.4

[a] From Intawongse, M. and Dean, J. R., *Food Adit. Contamin.*, **23**, 36–48 (2006).
[b] Mn and Zn measured by FAAS; Cu and Cd measured by DP-ASV.
[c] ND, non-detectable.

(16.5%). In contrast, manganese in the vegetables was extracted to the least extent under neutral-pH conditions (ca. 10–20%), i.e. in the intestinal fluid. The highest content of manganese in spinach (74.7%) was measured in the residual phase while there were significantly lower concentrations in lettuce (25.9%) and radish (33.3%).

7.2.3.4 Zinc

The major contents of zinc in lettuce (47.8%), spinach (87.7%) and radish (58.5%) remained in the residual fraction. Conversely, the least amounts (less than 10%) were solubilized in the intestinal phase. In the gastric phase, only 7% of zinc in spinach was extracted, while 45.2 and 32.3% were recovered from lettuce and radish, respectively.

7.2.4 Conclusions

The results indicate that metal bioavailability varied widely from element to element and according to the different plant types. The greatest extents of 'released metals' were found in lettuce (Mn, 63.7%), radish (Cu, 62.5%), radish (Cd, 54.9%), radish (Mn, 45.8%) and lettuce (Zn, 45.2%).

7.3 Bioavailability of POPs by Plants[§]

7.3.1 Background

This case study assesses the uptake of three persistent organic pollutants (POPs), namely, α-endosulfan, β-endosulfan and endosulfan sulfate in lettuce plants. The latter were grown on compost that had previously been 'spiked' at two concentration levels, i.e. 10 and 50 μg g^{-1} per POP. The plants were grown under artificial daylight for 12 h a day. Uptake was assessed by measuring the amount of endosulfan compounds in the roots and leaves from the lettuce plants after 10, 20 and 33 days. In addition, control plants grown in uncontaminated soil were monitored and analysed. All samples (soil and lettuce) were extracted using pressurized fluid extraction and analysed using gas chromatography with mass-selective detection.

7.3.2 Experimental

7.3.2.1 Reagents

Standard solutions were prepared in dichloromethane, with α-endosulfan (99.6 wt%), β-endosulfan (99.9 wt%) and endosulfan sulfate (97.7 wt%) provided by Riedel-de Haën (Steinheim, Germany). Pentachloronitrobenzene (PCNB) (99 wt%) was employed as the internal standard (Aldrich, Steinheim, Germany).

[§] From Esteve-Turrillas, F. A., Scott, W. R., Pastor, A. and Dean, J. R., *J. Environ. Monit.*, **7**, 1093–1098 (2006).

Acetone and dichloromethane were provided by Fisher Scientific UK Ltd (Loughborough, Leicestershire, UK) and anhydrous Na_2SO_4 by BDH Chemicals Ltd (Poole, UK).

7.3.2.2 Pressurized Fluid Extraction of Soil or Lettuce Samples

A Dionex Accelerated Solvent Extraction (ASETM 200) instrument (Sunnyvale, CA, USA) was used to extract the soil and lettuce samples. 'Hydromatrix', supplied by Varian Ltd (Walton-on-Thames, Surrey, UK) was employed for sample drying and sample dispersion during the extraction process. For soil extraction, ca. 8 g (accurately weighed) of soil and hydromatrix were added to a 33 ml cell. The full cell was closed and extracted at 100°C and 2000 psi, for 10 min, with acetone:dichloromethane 1:1 (vol/vol) as the solvent. This extract was evaporated under a nitrogen flow to <10 ml and then 20 μl of internal standard was added. The final extract solution (10.0 ml) was analysed by GC–MS. For lettuce extraction, ca. 5 g (accurately weighed) of lettuce and hydromatrix were placed in an 11 ml cell. Two sequential extractions were performed at 100°C and 2000 psi, over 10 min, with acetone:dichloromethane 1:1 (vol/vol) as the solvent. In order to remove the extracted water, 5 g of anhydrous Na_2SO_4 was added 'post-extraction'; this extract was evaporated to dryness under a stream of nitrogen and then 0.5 ml of an internal standard $(10 \mu g \, ml^{-1})$ was added and this solution analysed by GC–MS.

7.3.2.3 Preparation of Soil and Growing of Lettuce Plants

Contaminated soils at two levels of concentration were used for this study. For the low-level system, 2 l of acetone containing 8 mg of each endosulfan compound standard was added to 800 g of soil and mixed thoroughly. The high-level system was prepared in the same manner except in this case with 40 mg of endosulfan standard. The final endosulfan compound concentrations in soil were estimated to be 10 and 50 $\mu g \, g^{-1}$, for the low- and high-contamination levels, respectively. After air-drying for 48 h, the soils were mixed to ensure homogeneity prior to growing the lettuce seedling plants.

Small lettuce seedlings (ca. 10 cm in height and weighing 15–20 g) were transplanted into individual pots with 40 g of 'endosulfan-spiked' soil. Several lettuces were also planted in 'unspiked' soil as control samples. The lettuce plants were grown over 10, 20 and 33 days, under artificial light with time-intervals of 12 h daylight and 12 h in the dark. The air temperature and humidity were monitored over the growing duration. The air temperature was within the range 15.2 to 24.1°C while the humidity varied between 43 and 79%. The water retention capacity of the soil was experimentally determined to be a minimum of 20 ml and therefore each plant was watered daily with 20 ml of water. No excess water resulted from this process. α-Endosulfan, β-endosulfan and endosulfan sulfate determinations were then carried out on the lettuce plants (roots and leaves).

7.3.2.4 GC–MS Analysis

A Hewlett Packard gas chromatograph (HP G1800A GCD) (Palo Alto, CA, USA) with a Hewlett Packard HP-5 ms capillary column (30 m × 0.25 mm id, 0.25 μm film thickness), equipped with a quadrupole mass spectrometer detector, was used for POP determinations. An injection volume of 1 μl was employed in the split mode (1:4). The injector temperature was 250°C and helium was used as the carrier gas in the constant-flow mode at a rate of 1 ml min^{-1}. The temperature programme of the oven was as follows: 60°C, held for 1 min, increased at a rate of 15°C min^{-1} to 180°C, then at a second rate of 3°C min^{-1} to 250°C and finally held for 1 min. The detector temperature was 280°C and measurements were carried out in the selected-ion-monitoring (SIM) acquisition mode.

7.3.3 Results and Discussion

The uptakes of endosulfan compounds grown on contaminated soil (10 and 50 μg g^{-1}) were determined. Control lettuce plants were also grown on unadulterated soil. Lettuce plants were harvested after 10, 20 and 33 days and the results are shown in Table 7.6. In accordance with expectations, it is noted that endosulfan compounds are detected in the roots of lettuce plants even after 10 days. However, a growing period of 33 days on the most contaminated soil (50 μg g^{-1}) is required before any endosulfan compounds are detected in the leaves. The root-to-leaf ratio was found to be 3.1 for α-endosulfan, 46.0 for β-endosulfan and 24.3 for endosulfan sulfate. It was also noted that during the growing period between 10 and 33 days the amounts of endosulfan compounds in the roots of the lettuce plants increase.

Table 7.6 Uptake of endosulfan compounds by lettuce plants grown on contaminated soil, followed by PFE-GC–MS[a,b]

Endosulfan concentration in soil (μg g^{-1})	Compound	Endosulfan concentration (μg g^{-1})[c]					
		Day 10		Day 20		Day 33	
		Root	Leaf	Root	Leaf	Root	Leaf
10	α-Endosulfan	0.3 ± 0.2	< LOD	1.4 ± 0.7	< LOD	1.7 ± 0.3	< LOD
	β-Endosulfan	0.12 ± 0.08	< LOD	1.0 ± 0.6	< LOD	1.5 ± 0.3	< LOD
	Endosulfan sulfate	0.14 ± 0.06	< LOD	1.2 ± 0.5	< LOD	1.5 ± 0.4	< LOD
50	α-Endosulfan	0.5 ± 0.3	< LOD	1.7 ± 0.8	< LOD	2.5 ± 0.6	0.8 ± 0.004
	β-Endosulfan	0.4 ± 0.2	< LOD	0.8 ± 0.3	< LOD	2.3 ± 0.6	0.05 ± 0.01
	Endosulfan sulfate	0.5 ± 0.2	< LOD	1.2 ± 0.4	< LOD	1.7 ± 0.4	0.07 ± 0.01

[a] < LOD, less than the limit of detection.
[b] From Esteve-Turrillas, F. A., Scott, W. C., Pastor, A. and Dean, J. R., *J. Environ. Monit.*, **7**, 1093–1098 (2005).
[c] Mean ± SD, $n = 3$.

7.3.4 Conclusions

Endosulfan compounds were found in the roots of all lettuce plants, irrespective of the soil-'spike' level or age of the plant. Endosulfan compounds were found in the leaves of the '33-day' lettuce plants (soil 'spiked' at $50 \mu g g^{-1}$). The root-to-leaf ratio was found to be 3.1 for α-endosulfan, 46.0 for β-endosulfan and 24.3 for endosulfan sulfate.

7.4 Bioaccessibility of POPs from Plants[¶]

7.4.1 Background

This case study assesses the oral bioaccessibility of three persistent organic pollutants (POPs), namely, α-endosulfan, β-endosulfan and endosulfan sulfate, from 'spiked' lettuce plants. Oral bioaccessibility was assessed using a simulated *in vitro* gastrointestinal extraction technique.

All aqueous samples were extracted using liquid–liquid extraction while (semi)-solid samples were extracted using pressurized fluid extraction and analysed by using gas chromatography with mass-selective detection.

7.4.2 Experimental

7.4.2.1 Reagents

Standard solutions were prepared in dichloromethane (DCM), with α-endosulfan (99.6 wt%), β-endosulfan (99.9 wt%) and endosulfan sulfate (97.7 wt%) provided by Riedel-de Haën (Steinheim, Germany). Pentachloronitrobenzene (PCNB) (99 wt%) was employed as the internal standard (Aldrich, Steinheim, Germany). Acetone and dichloromethane were provided by Fisher Scientific UK Ltd (Loughborough, Leicestershire UK) and anhydrous Na_2SO_4 by BDH Chemicals Ltd (Poole, Dorset, UK). Pepsin-A powder (1 anson unit per g) (lactose as diluent) and amylase were provided by BDH Chemicals Ltd (Poole, Dorset, UK), bile salts from Sigma Chemicals (Poole, Dorset, UK) and pancreatin from Fisher Scientific UK Ltd (Loughborough, Leicestershire, UK).

7.4.2.2 Liquid–Liquid Extraction of Water, Gastric and Intestinal Juice

Aqueous samples were extracted by using 3×10 ml of DCM. The samples consisted of distilled water (15 ml), gastric juice (0.1% (wt/vol) pepsin in saline) and intestinal juice (3% (wt/vol) pancreatin, 1% (wt/vol) amylase and 0.15% (wt/vol) bile salts). Each sample was 'spiked' with endosulfan compounds (all at concentrations of $10 \mu g ml^{-1}$) to assess the influence of the sample matrix on recovery. After addition of the internal standard (25 ppm), the extracts were analysed by using GC–MS.

[¶] From Esteve-Turrillas, F. A., Scott, W. R., Pastor, A. and Dean, J. R., *J. Environ. Monit.*, **7**, 1093–1098 (2006).

7.4.2.3 In vitro *Gastrointestinal Extraction of Lettuce*

This determination consists of two sequential processes, namely, a gastric and an intestinal digestion, with each one carried out under simulated human conditions (enzymes, pH and temperature). In the first stage, ca. 5 g (accurately weighed) of lettuce (chopped into several pieces) was treated with 15 ml of pepsin (0.1% (wt/vol) in saline). The pH of the solution was adjusted to pH 1.8 with diluted HCl. The mixture was then shaken at 100 rpm in a thermostatic bath maintained at 37°C. The pH was measured every 30 min and maintained below 2.5 by using HCl. After 3 h, the solution was centrifuged at 3000 rpm for 5 min and then filtered.

The second stage involved extraction with intestinal juices. To the gastric digest residue, 5 ml of pancreatin (3% (wt/vol)), 5 ml of amylase (1% (wt/vol)) and 5 ml of bile salts (0.15% (wt/vol)), all in saline solution, were added. Diluted NaOH was used to maintain the pH at ~7 while shaking at 100 rpm for 3 h in a thermostatic bath, employing the same conditions as before. All extracts (gastric and intestinal) were extracted with 3×15 ml of DCM. These were then evaporated to dryness under a nitrogen flow and 0.5 ml of an internal standard ($10 \mu g \, ml^{-1}$) was added prior to GC–MS analysis. The resultant sample residues were analysed by PFE.

7.4.2.4 Pressurized Fluid Extraction of Lettuce Sample Residue

A Dionex Accelerated Solvent Extraction (ASE™ 200) instrument (Sunnyvale, CA, USA) was used to extract the soil and lettuce samples. 'Hydromatrix', supplied by Varian Ltd (Walton-on-Thames, Surrey, UK) was employed for sample drying and sample dispersion during the extraction process. Approximately 5 g (accurately weighed) of lettuce sample residue and hydromatrix were placed in a 11 ml cell. Two sequential extractions were performed at 100°C and 2000 psi, over 10 min, with acetone:dichloromethane 1:1 (vol/vol) as solvent. In order to remove the extracted water, 5 g of anhydrous Na_2SO_4 was added 'post-extraction'; this extract was evaporated to dryness under a stream of nitrogen and then 0.5 ml of an internal standard ($10 \mu g \, ml^{-1}$) was added and this solution analysed by GC–MS.

7.4.2.5 GC–MS Analysis

A Hewlett Packard gas chromatograph (HP G1800A GCD) (Palo Alto, CA, USA) with a Hewlett Packard HP-5 ms capillary column (30 m $\times$ 0.25 mm id, 0.25 μm film thickness), equipped with a quadrupole mass spectrometer detector, was used for POP determinations. An injection volume of 1 μl was employed in the split mode (1:4). The injector temperature was 250°C and helium was used as the carrier gas in the constant-flow mode at a rate of 1 ml min^{-1}. The temperature programme of the oven was as follows: 60°C, held for 1 min, increased at a rate of 15°C min^{-1} to 180°C, then at a second rate of 3°C min^{-1} to 250°C and finally held for 1 min. The detector temperature was 280°C and measurements were carried out in the selected-ion-monitoring (SIM) acquisition mode.

7.4.3 Results and Discussion

The *in vitro* gastrointestinal extraction approach consists of two procedures which simulate human digestion. The first stage involves extraction with gastric juice (pepsin) to simulate the activity of the stomach, while stage two involves extraction with intestinal juice (pancreatin, amylase and bile salts) which simulates the activity of the intestines. These extractions were performed in a thermostatic bath, at 37°C, with agitation over several hours. After this, liquid–liquid extractions of the gastric and intestinal extracts were carried out to determine the recovery of endosulfan compounds. Recovery tests of these processes were performed in order to assess the potential for decomposition and losses of some endosulfan compounds due to the presence of enzyme(s) and the influence of pH. Table 7.7 shows the results obtained for the recoveries of endosulfan compounds by liquid–liquid extraction in gastric and intestinal juices. The extraction of endosulfan compounds from distilled water was carried out as a control to enable the effect of the *in vitro* gastrointestinal extraction process to be investigated. It is noted (Table 7.7) that the recoveries in the gastric and intestinal juices ranged from 82.5 to 92.1%, which is comparable to the recoveries of endosulfan compounds from distilled water (88.4–92.4%).

Assessment of the potential of gastrointestinal extraction to determine the bioavailability of endosulfan compounds required the use of 'spiked' lettuce leaves. Lettuce leaves were 'spiked' at a concentration of $10 \, \mu g \, g^{-1}$. The lettuce leaves were then subjected to *in vitro* gastrointestinal extraction. The resultant lettuce residues were then extracted using PFE and the results obtained are shown in Table 7.8. These results indicate that the bioavailability of endosulfan compounds in spiked lettuce leaves is minimal (< 3.5% bioavailability). The majority of the

Table 7.7 Liquid–liquid extraction of endosulfan compounds, at a spiking level of $10 \, \mu g \, ml^{-1}$, from aqueous solution, gastric juice and intestinal juice, followed by PFE-GC–MS[a]

Compound	Recovery based on three sequential extractions (mean % ± SD, $n = 5$)		
	Water	Gastric juice at pH 2.5	Intestinal juice at pH 7
α-Endosulfan	89.9 ± 0.12	82.1 ± 0.21	85.0 ± 0.20
β-Endosulfan	88.5 ± 0.19	86.1 ± 0.19	83.5 ± 0.22
Endosulfan sulfate	88.6 ± 0.34	85.4 ± 0.33	85.1 ± 0.20

[a] From Esteve-Turrillas, F. A., Scott, W. C., Pastor, A. and Dean, J. R., *J. Environ. Monit.*, **7**, 1093–1098 (2005).

Table 7.8 Recovery of endosulfan compounds from spiked lettuce leaves, at a spiking level of $10 \,\mu g \, g^{-1}$, using PFE-GC–MS and an assessment of the bioavailability of endosulfan compounds using simulated *in vitro* gastrointestinal extractions followed by LLE-GC–MS[a]

Compound	Recovery following *in vitro* gastric extraction $(\mu g \, g^{-1})^{b,c,d}$	Recovery following *in vitro* intestinal extraction $(\mu g \, g^{-1})^{b,c}$	Residual fraction $(\mu g \, g^{-1})^{b,e,f}$	Total recovery after gastro-intestinal extractions plus residual fraction $(\mu g \, g^{-1})$
α-Endosulfan	ND	0.06 ± 0.02	3.51 ± 0.64	3.56
β-Endosulfan	ND	0.12 ± 0.02	3.69 ± 0.76	3.81
Endosulfan sulfate	ND	0.09 ± 0.01	3.35 ± 0.72	3.44

[a] From Esteve-Turrillas, F. A., Scott, W. C., Pastor, A. and Dean, J. R., *J. Environ. Monit.*, **7**, 1093–1098 (2005).
[b] Mean $\pm$ SD, $n = 3$.
[c] Measured by LLE-GC–MS.
[d] ND, non-detectable.
[e] Reported on a dry-weight basis.
[f] Measured by PFE-GC–MS.

endosulfan compounds were not extracted by the *in vitro* gastrointestinal extraction approach, but remained within the lettuce matrix and were subsequently recovered by PFE. The total recovery by *in vitro* gastrointestinal extraction and PFE of the residual fraction amounted to ca. 35%. This non-quantitative recovery was attributed to losses caused by the transfer of lettuce material from the liquid–liquid extraction stage to the PFE stage.

The simulated *in vitro* gastric intestinal extraction was based on 3 h gastric and 3 h intestinal extraction. In the case of gastric extraction simulating activity in the stomach, this represents the upper limit for food to be present. Typically, food is retained in the stomach from between 8 min to 3 h. In the case of intestinal extraction, food samples can be present in the duodenum, jejunum and ileum for 30 min, 1.5 h and 4–5 h, respectively. Therefore, in this procedure, lettuce samples are subjected to approximately 50% of the potential time for intestinal absorption. Thus, in principle, the levels of intestinal extraction could double (increase by 50%) which would equate to $< 7\%$ bioavailability ($0.034 \,\mu g \, g^{-1}$ for α-endosulfan, $0.065 \,\mu g \, g^{-1}$ for β-endosulfan, and $0.054 \,\mu g \, g^{-1}$ for endosulfan sulfate).

7.4.4 Conclusions

An assessment of the bioavailability of endosulfan compounds using *in vitro* gastrointestinal extraction was carried out. It was found that minimal amounts of endosulfan compounds were determined to be available for absorption in the gut, based on the simulated extraction procedure, i.e. no detectable endosulfan compounds were determined in the gastric extracts, while small quantities (ranging from 0.06 to 0.12 μg g^{-1}) were found in the intestinal extraction.

Summary

In this chapter, four representative examples of case studies (all taken from the author's own laboratory) have been provided in order to illustrate the 'typical' approaches that need to be adopted in the investigation of bioavailability and (oral) bioaccessibility of metals and persistent organic pollutants from plants and soil samples.

Chapter 8

Recording of Information and Selected Resources

Learning Objectives

- To enable relevant scientific information to be recorded in the laboratory.
- To allow correct sample pre-treatment details to be noted.
- To allow correct sample preparation details to be noted.
- To allow correct analytical technique information to be noted.
- To suggest relevant resources that can be consulted, including books, journals and the Internet.

8.1 Safety

Prior to commencing any experimentation, you are required to complete a hazard and risk assessment of the chemicals and apparatus that you will use, i.e. a Control of Substances Hazardous to Health (COSHH) assessment.

The key safety points when handling chemicals in the laboratory are as follows:

- Treat all chemicals as being potentially dangerous.

- Wear a laboratory coat, with the buttons fastened, at all times.

- Wear eye protection at all times.

- Make sure that you know where safety equipment, such as an eye bath, fire extinguisher and first-aid kit, are kept before you start work.

Bioavailability, Bioaccessibility and Mobility of Environmental Contaminants   John R. Dean
© 2007 John Wiley & Sons, Ltd

- Wear gloves for toxic, irritant or corrosive chemicals.

- Perform procedures with toxic, irritant or corrosive chemicals in a fume cupboard.

- Label all solutions/samples appropriately.

- Extinguish all naked flames when working with flammable substances.

- Never drink, eat or smoke when chemicals are being handled.

- Report all spillages and clean them up appropriately.

- Dispose of chemicals in the correct manner.

8.2 Recording of Information in the Laboratory

8.2.1 Introduction

When carrying out any laboratory work, it is important to record information in a systematic manner relating to sample information, sample treatment and analytical technique used, calibration strategy and recording of results. Included below are some examples of data sheets that could be used to ensure that all information is recorded in a systematic manner. These data sheets are not intended to be totally comprehensive, and so may be altered and amended as required.

The first data sheet (A) allows identification of any preliminary sample pretreatment that may be necessary prior to sample preparation. Data sheets (B) and (C) identify the sample preparation techniques for inorganic and organic samples, respectively, while data sheets (D) and (E) contain, respectively, the instrumental requirements for the most commonly used inorganic and organic analytical techniques. (Data sheet (A) is taken from Dean, J. R., *Practical Inductively Coupled Plasma Spectroscopy*, AnTS Series. Copyright 2005. © John Wiley and Sons, Limited. Reproduced with permission. Data sheets (B), (C), (D) and (E) are taken from Dean, J. R., *Methods for Environmental Trace Analysis*, AnTS Series. Copyright 2003. © John Wiley and Sons, Limited. Reproduced with permission).

8.2.2 Examples of Data Sheets

Data Sheet A: Sample Pre-Treatment

- Sample description/'unique' identifier ....

- Source/location of sample ....

- Sample dried   yes/no
  Oven-dried (temperature and duration) ....°C/h
  Air-dried (temperature and duration) ....°C/h
  Other (specify)..

- Grinding and sieving
  Grinder used (model/type)...
  Particle size (sieve mesh size)..

- Mixing of the sample
  Manual shaking   yes/no
  Mechanical shaking   yes/no
  Other (specify)..

- Sample storage
  Fridge   yes/no
  Other (specify)..

- Chemical pre-treatment
  pH adjustment   yes/no
  Addition of alkali (specify) .... or acid (specify) ....
  or buffer .... pH = ...

Data Sheet B: Sample Preparation for Inorganic Analysis

- Sample weight(s)
  (record to four decimal places)

  Sample 1 ....g   Sample 2 ....g
- Acid-digestion
  Vessel used...
  Hot-plate or other (specify)..
  Temperature-controlled or not (specify) ....
  Type of acid(s) used...
  Volume of acid used ....ml
  Duration of digestion... h/min
  Any other details ....
- Other methods of sample decomposition, e.g. fusion, dry ashing (specify)
  ....
  ....
  ....

- Sample derivatization
  Specify ....
  ....
  Reagent concentration .... mol l^{-1}
  Reagent volume used ....ml
  Heat required (specify) ....
- Sample dilution
  Specify the dilution factor involved, with appropriate units...............
  ....
  ....
  ....

- Addition of an internal standard
  Specify ....
  Added before digestion   yes/no
  Added after digestion   yes/no
- Sample and reagent blanks
  Specify ....
  ....
- Recovery
  Specify ....
  Added before digestion   yes/no
  Added after digestion   yes/no

Data Sheet C: Sample Preparation for Organic Analysis

- Sample weight(s)
  (record to four decimal places)
  Sample 1 ....g   Sample 2 .... g
- Soxhlet extraction method
  Drying agent added and weight (specify)...
  Type of solvent(s) used...
  Volume of solvent(s) used.. ml
  Duration of extraction ....h/min
  Any other details...
- Other method of sample extraction e.g. SFE, ASE (PFE), SPME or MAE, plus
  operating conditions (specify)
  ....
  ....
  ....
- Sample clean-up   yes/no
  Specify ....
  ....
- Pre-concentration of the sample   yes/no
  Method of solvent reduction ....
  Final volume of extract..
- Sample derivatization
  Specify ....
  ....
  Reagent concentration .... $mol\,l^{-1}$
  Reagent volume used .... ml
  Heat required (specify) ....
- Sample dilution
  Specify the dilution factor involved, with appropriate units ....
  ....
  ....
  ....
- Addition of an internal standard
  Specify ....
  Added before extraction   yes/no
  Added after extraction   yes/no
- Sample and reagent blanks
  Specify ....
  ....
- Recovery
  Specify ....
  Added before extraction   yes/no
  Added after extraction   yes/no

Data Sheet D: Analytical Techniques for Inorganic Analysis

Method
   ICP–AES   yes/no
   ICP–MS   yes/no
   FAAS   yes/no
   GFAAS   yes/no
   Flame photometry   yes/no
   XRF   yes/no
   Other (specify) ....

Inductively Coupled Plasma–Atomic Emission Spectroscopy

- ICP characteristics
  Manufacturer ....
  Frequency .... Hz
  Power ....kW
  Observation height .... mm above load coil

- Argon gas flow rates
  Outer gas flow rate ....$1\,min^{-1}$
  Intermediate gas flow rate ....$1\,min^{-1}$
  Injector gas flow rate ....$1\,min^{-1}$

- Sample introduction method
  Nebulizer/spray chamber (specify) ....
  ....
  ....

- Spectrometer
  Simultaneous or sequential
  Element(s) and wavelength(s) ....
  ....

- Quantitation
  Peak height   yes/no
  Peak area   yes/no
  Method used   manual/electronic
  Internal standard ....
  External standard ....
  Calibration method   direct/standard additions
  Number of calibration standards ....
  Linear range of calibration ....

Inductively Coupled Plasma–Mass Spectrometry

- ICP characteristics
  Manufacturer ....
  Frequency .... Hz
  Power ....kW

- Argon gas flow rates
  Outer gas flow rate .... $1 \, min^{-1}$
  Intermediate gas flow rate .... $1 \, min^{-1}$
  Injector gas flow rate .... $1 \, min^{-1}$

- Sample introduction method
  Nebulizer/spray chamber (specify) ....
  ....
  ....

- Mass spectrometer
  Element(s) and mass/charge ratio(s) ....
  ....
  Scanning or peak hopping mode

- Collision/reaction cell .... yes/no
  Gas(es) used ....
  Flow rate .... $ml \, min^{-1}$
  Used for which elements? ....

- Quantitation
  Peak height   yes/no
  Peak area   yes/no
  Method used   manual/electronic
  Internal standard ....
  External standard ....
  Calibration method   direct/standard additions
  Number of calibration standards ....
  Linear range of calibration ....

Flame Atomic Absorption Spectroscopy

- Atomizer characteristics
  Flame .... Graphite furnace ....
  Cold-vapour .... Hydride generation ....

- Flame – gas mixture ....
  Gas flow rates .... $1 \, min^{-1}$

- Sample introduction method
  Nebulizer/expansion chamber (specify) ....
  ....
  ....

- spectrometer
  Element(s) and wavelength(s) ....
  ....

- Quantitation
  Peak height   yes/no
  Peak area   yes/no
  Method used   manual/electronic
  Calibration method   direct/standard additions
  Number of calibration standards ....
  Linear range of calibration ....

Electrothermal (Graphite-Furnace) Atomic Absorption Spectroscopy

- Atomizer characteristics
  graphite furnace ....

- Furnace characteristics
  Graphite .... Pyrolytic coated ....
  L'Vov platform .... Other (specify) ....

- Programme
  Drying .... temperature   °C; time   s
  Ashing .... temperature   °C; time   s
  Atomization .... temperature   °C; time   s
  Other..................... temperature   °C; time   s

- Background correction   yes/no
  Deuterium   yes/no
  Zeeman   yes/no
  Smith–Hieftje   yes/no

- Spectrometer
  Element(s) and wavelength(s) ....
  ....

- Quantitation
  Peak height   yes/no
  Peak area   yes/no
  Method used   manual/electronic
  Calibration method   direct/standard additions
  Number of calibration standards ....
  Linear range of calibration ....

Flame Atomic Emission Spectroscopy (Flame Photometry)

- Atomizer characteristics
  Flame ....

- Flame – gas mixture ....
  Gas flow rates ....

- Sample introduction method
  Nebulizer/expansion chamber (specify) ....
  ....
  ....

- Spectrometer
  Element(s)...

- Quantitation
  Peak height   yes/no
  Method used   manual/electronic
  Calibration method   direct/standard additions
  Number of calibration standards ....
  Linear range of calibration ....

X-Ray Fluorescence Spectroscopy

- Spectrometer
  Manufacturer ....
  Energy/wavelength-dispersive ....

- Sample preparation
  Specify ....
  ....
  ....

- Quantitation
  Peak height   yes/no
  Peak area   yes/no
  Internal standard ....
  External standard ....
  Calibration method   direct/standard additions
  Number of calibration standards ....
  Linear range of calibration ....

Data Sheet E: Analytical Techniques for Organic Analysis

Chromatographic Method

> GC   yes/no
> HPLC   yes/no
> IC   yes/no
> Other (specify) ....

Gas Chromatography

- Column characteristics
  Manufacturer ....
  Type ....
  Length ....m;   internal diameter .... mm
  Film thickness ....μm

- Carrier gas and flow rate ....

- Isothermal or temperature-programmed   yes/no
  Temperature programme (specify) ....
  ....
  ....

- Injector type ....
  Injector temperature ....°C
  Split ratio ....
  Injection volume ....μl

- Detector type ....
  Operating conditions ....

- Quantitation
  Peak height   yes/no
  Peak area   yes/no
  Method used   manual/electronic
  Internal standard ....
  External standard ....
  Calibration method   direct/standard additions
  Number of calibration standards ....
  Linear range of calibration ....

High Performance Liquid Chromatography

- Column characteristics
  Manufacturer ....
  Type ....
  Reversed-phase or normal-phase ....
  Length .... cm; internal diameter ....mm

- Mobile phase (solvent) ....
  Flow rate .... ml min^{-1}

- Isocratic or gradient   yes/no
  Gradient programme (specify) ....
  ....
  ....

- Injector volume .... µl

- Detector type ....
  Operating conditions ....

- Quantitation
  Peak height   yes/no
  Peak area   yes/no
  Method used   manual/electronic
  Internal standard ....
  External standard ....
  Calibration method: direct/standard additions
  Number of calibration standards: ....
  Linear range of calibration ....

8.3 Selected Resources

To assist readers to maximize their time, and hence efficiency, a selected resource list is provided.

The most appropriate approach to keep up-to-date is to consult relevant journals which publish research results on the use and application of analytical techniques. Some suggested journals are given in Table 8.1.

It is now common practice to find that publishers allow their journal contents to be accessed electronically, often for free (no charge). By 'electronically' signing up for their e-mail 'alerting service', you can automatically be sent the latest contents pages for each journal selected. This allows the contents of each selected journal to be automatically forwarded to you directly at your e-mail address, thus enabling you to view the latest publications in a particular journal as they are published. This can be a valuable tool in obtaining the latest research information. A selection of major publishers of journals, plus their corresponding websites, is presented in Table 8.2.

Most journals are also available electronically on your desktop PC. This allows the full text to be read in either PDF or HTML formats. In PDF format, the article appears exactly like the print copy that you might find on a library shelf. In HTML format, the article will have hyperlinks to tables, figures or references (the references may be further linked to their original source using a 'reference linking' service). However, this facility requires the payment of a subscription fee, as is often the case for libraries in universities, industry or public organizations, etc.

Some specific books that may be useful for further in-depth study of the field of bioavailability and bioaccessibility are shown in Table 8.3. Other books that cover related issues, such as sampling and sample preparation, are shown in Tables 8.4 and 8.5, respectively.

An alternative source of material is the Internet or World Wide Web. However, caution needs to be exercised in the use of the Internet in terms of the quality of the source material. It is recommended that only sources of known repute (e.g. professional bodies) are used. Some examples of suitable websites are given in Table 8.6.

Table 8.1 Alphabetical list of selected journals that publish articles (research papers, communications, critical reviews, etc.) on analytical techniques and their applications[a]

Journal	Publisher	Web address[b]
Analyst	The Royal Society of Chemistry	http://www.rsc.org/Publishing/Journals/an/
Analytica Chimica Acta	Elsevier	http://www.elsevier.com/wps/find/journaldescription.cws_home/502681/description#description
Analytical Chemistry	American Chemical Society	http://pubs.acs.org/journals/ancham/index.html
Chemosphere	Elsevier	http://www.elsevier.com/wps/find/journaldescription.cws_home/362/description#description
Environment International	Elsevier	http://www.elsevier.com/wps/find/journaldescription.cws_home/326/description#description
Environmental Pollution	Elsevier	http://www.elsevier.com/wps/find/journaldescription.cws_home/405856/description#description
Environmental Science and Technology	American Chemical Society	http://pubs.acs.org/journals/esthag/
Journal of Agricultural and Food Chemistry	American Chemical Society	http://pubs.acs.org/journals/jafcau/index.html
Journal of Analytical Atomic Spectrometry (JAAS)	The Royal Society of Chemistry	http://www.rsc.org/Publishing/Journals/JA/index.asp
Journal of Chromatography (Parts A and B)	Elsevier	http://www.elsevier.com/wps/find/journaldescription.cws_home/502688/description#description
Journal of Environmental Monitoring (JEM)	The Royal Society of Chemistry	http://www.rsc.org/Publishing/Journals/em/index.asp
Microchemical Journal	Elsevier	http://www.elsevier.com/wps/find/journaldescription.cws_home/620391/description#description
Science of the Total Environment	Elsevier	http://www.elsevier.com/wps/find/journaldescription.cws_home/503360/description#description
Soil Biology and Biochemistry	Elsevier	http://www.elsevier.com/wps/find/journaldescription.cws_home/332/description#description
Spectrochimica Acta (Part B)	Elsevier	http://www.elsevier.com/wps/find/journaldescription.cws_home/525437/description#description
Talanta	Elsevier	http://www.elsevier.com/wps/find/journaldescription.cws_home/525438/description#description
Trends in Analytical Chemistry	Elsevier	http://www.elsevier.com/wps/find/journaldescription.cws_home/502695/description#description

[a] Adapted from Dean, J. R., *Practical Inductively Coupled Plasma Spectroscopy*, AnTS Series, John Wiley & Sons, Ltd, Chichester, UK, 2005.
[b] As of August 2006. The products or material displayed are not endorsed by the author or the publisher of this present text.

Table 8.2 Selected major publishers of journals. From Dean, J. R., *Practical Inductively Coupled Plasma Spectroscopy*, AnTS Series. Copyright 2005. © John Wiley and Sons, Limited. Reproduced with permission

Publisher	Web address[a]
The Royal Society of Chemistry	http://www.rsc.org
American Chemical Society	http://www.pubs.acs.org
Wiley	http://www.wiley.com
Elsevier	http://www.elsevier.com
CRC Press	http://www.crcpress.com
Springer-Verlag	http://www.springerlink.metapress.com

[a] As of August 2006. The products or material displayed are not endorsed by the author or the publisher of this present text.

Table 8.3 Specific books on Speciation, Fractionation, Bioavailability and Bioaccessibility[a,b]

Cornelis, R., Caruso, J. A., Crews, H. and Heumann, K. K. (Eds), *Handbook of Elemental Speciation*, Vol. 1, *Techniques and Methodology*, John Wiley & Sons, Ltd, Chichester, UK, 2005.

Cornelis, R., Caruso, J. A. Crews, H. and Heumann, K. G. (Eds), *Handbook of Elemental Speciation*, Vol. 2, *Species in the Environment, Food, Medicine and Occupational Health,* John Wiley & Sons, Ltd, Chichester, UK, 2005.

Szpunar, J. and Lobinski, R., *Hyphenated Techniques in Speciation Analysis*, The Royal Society of Chemistry, Cambridge, UK, 2004.

Cornelius, R, Caruso, J. A., Crews, H. and Heumann, K. G. (Eds), *Handbook of Elemental Speciation: Techniques and Methodology*, John Wiley & Sons, Ltd, Chichester, UK, 2003.

National Research Council, *Bioavailability of Contaminants in Soils and Sediments. Processes, Tools and Applications*, National Research Council, The National Academies Press, Washington, DC, USA, 2003.

Ebdon, L. Pitts, L., Cornelius, R., Crews, H., Donard, A. F. X. and Quevauviller, Ph., *Trace Element Speciation for Environment, Food and Health*, The Royal Society of Chemistry, Cambridge, UK, 2002.

Quevauviller, Ph., *Methodologies for Soil and Sediment Fractionation Studies*, The Royal Society of Chemistry, Cambridge, UK, 2002.

Caruso, J. A., Sutton, K. L. and Ackley, K. L. (Eds), *Elemental Speciation,* Elsevier, Amsterdam, The Netherlands, 2000.

Baveye, P., Block, J.-C. and Goncharuk, V V. (Eds) *Bioavailability of Organic Xenobiotics in the Environment: Practical Consequences for the Environment*, Springer-Verlag, Berlin, Germany, 1999.

(*continued overleaf*)

Table 8.3 (*continued*)

Quevauviller, Ph., *Method Performance Studies for Speciation Analysis*, The Royal Society of Chemistry, Cambridge, UK, 1997.

Van der Sloot, H. A., Heasman, L. and Quevauviller, Ph., *Harmonization of Leaching/Extraction Tests*, Elsevier, Amsterdam, The Netherlands, 1997.

Caroli, S., *Element Speciation in Bioinorganic Chemistry*, John Wiley & Sons, Ltd, Chichester, UK, 1996.

Ure, A. M. and Davidson, C. M., *Chemical Speciation in the Environment*, Blackie Academic and Professional, Glasgow, Scotland, UK, 1995.

Kramer, J. R. and Allen, H. E. *Metal Speciation. Theory, Analysis and Application*, Lewis Publishers Inc., Chelsea, MI, USA, 1988.

[a] Adapted from Dean, J. R., *Practical Inductively Coupled Plasma Spectroscopy*, AnTS Series, John Wiley & Sons, Ltd, Chichester, UK, 2005.
[b] Arranged in chronological order.

Table 8.4 Specific books on Sampling.[a] From Dean, J. R., *Practical Inductively Coupled Plasma Spectroscopy*, AnTS Series. Copyright 2005. John Wiley and Sons, Limited. Reproduced with permission

Conklin, A. R., *Field Sampling: Principles and Practices in Environmental Analysis*, Marcel Dekker, New York, NY, USA, 2004.

Bodger, K., *Fundamentals of Environmental Sampling*, Rowman & Littlefield Publishers, Lanham, MD, USA, 2003.

Muntau, H. *Comparative Evaluation of European Methods for Sampling and Sample Treatment of Soil*, Diane Publishing Company, Collingdale, PA, USA, 2003.

Popek, E.P., *Sampling and Analysis of Environmental Chemical Pollutants: A Complete Guide*, Academic Press, Oxford, UK, 2003.

Csuros, M., *Environmental Sampling and Analysis for Metals*, CRC Press, Boca Raton, FL, USA, 2002.

Csuros, M., *Environmental Sampling and Analysis: Laboratory Manual*, CRC Press, Boca Raton, FL, USA, 1997.

Ostler, N.K., *Sampling and Analysis*, Prentice Hall, New York, NY, USA, 1997.

Hess, K., *Environmental Sampling for Unknowns*, CRC Press, Boca Raton, FL, USA, 1996.

Keith, L.H., *Principles of Environmental Sampling*, Oxford University Press, Oxford, UK, 1996.

Keith, L.H., *Compilation of EPA's Sampling and Analysis Methods*, CRC Press, Boca Raton, FL, USA, 1996.

Harsham, K.D., *Water Sampling for Pollution Regulation*, Taylor and Francis, London, UK, (1995).

Table 8.4 (*continued*)

Quevauviller, Ph., *Quality Assurance in Environmental Monitoring: Sampling and Sample Pretreatment*, John Wiley & Sons, Ltd, Chichester, UK, 1995.

Russell Boulding, J., *Description and Sampling of Contaminated Soils: A Field Guide*, CRC Press, Boca Raton, FL, USA, 1994.

Carter, M.R., *Soil Sampling and Methods of Analysis*, CRC Press, Boca Raton, FL, USA, 1993.

Baiuescu, G.E., Dumitrescu, P. and Gh. Zugravescu, P., *Sampling*, Ellis Horwood, London, UK, 1991.

Keith, L.H., *Environmental Sampling and Analysis: A Practical Guide*, Lewis Publishers Inc., Chelsea, MI, USA, 1991.

---

[a] Arranged in chronological order.

Table 8.5 Specific books on Sample Preparation[a,b]

Mester, Z and Sturgeon, R., *Sample Preparation for Trace Element Analysis*, Elsevier, Amsterdam, The Netherlands, 2004.

Dean, J. R., *Methods for Environmental Trace Analysis*, AnTS Series, John Wiley & Sons, Ltd, Chichester, UK, 2003.

Mitra, S. (Ed.), *Sample Preparation Techniques in Analytical Chemistry*, Wiley-Interscience, New York, NY, USA, 2003.

Dean, J. R., *Extraction Methods for Environmental Analysis*, John Wiley & Sons, Ltd, Chichester, UK, 1998.

Kingston, H. M. and Jassie, L. B., *Introduction to Microwave Sample Preparation*, ACS Professional Reference Book, American Chemical Society, Washington, DC, USA, 1988.

Bock, R., *A Handbook of Decomposition Methods in Analytical Chemistry*, International Textbook Company, London, UK, 1979.

---

[a] Adapted from Dean, J. R., *Practical Inductively Coupled Plasma Spectroscopy*, AnTS Series, John Wiley & Sons, Ltd, Chichester, UK, 2005.
[b] Arranged in chronological order.

Table 8.6 Selected useful web sites. From Dean, J. R., *Practical Inductively Coupled Plasma Spectroscopy*, AnTS Series. Copyright 2005. © John Wiley and Sons, Limited. Reproduced with permission

Organization	Web address[a]
American Chemical Society (ACS)	http://www.acs.org
International Union of Pure and Applied Chemistry (IUPAC)	http://iupac.chemsoc.org
Laboratory of the Government Chemist (LGC)	http://www.lgc.co.uk
National Institute of Standards and Technology (NIST) Laboratory	http://www.cstl.nist.gov

(*continued overleaf*)

Table 8.6 (*continued*)

Organization	Web address[a]
National Institute of Standards and Technology (NIST) 'WebBook'	http://webbook.nist.gov
The Royal Society of Chemistry (RSC)	http://www.rsc.org
Society of Chemical Industry (SCI)	http://sci.mond.org
United States Environmental Protection Agency (USEPA)	http://www.epa.gov
United States National Library of Medicine (USNLM)	http://chem.sis.nlm.nih.gov

[a] As of August 2006. The products or material displayed are not endorsed by the author or the publisher of this present text.

Summary

Accurate recording of experimental details prior to, during and after practical work is an essential component in analytical science. This chapter has attempted to capture the essential information which needs to be recorded at the same time as the practical work is being undertaken. Templates (data sheets) have been provided, as guidance, in order to ensure that the most appropriate information is recorded. In addition, details are given on the range of resources available, in both print and electronic format, to assist in the understanding of bioavailability and bioaccessibility of environmental contaminants.

Responses to Self-Assessment Questions

Chapter 1

Response 1.1

Now, while local councils in the UK operate waste recycling schemes, a considerable amount of the waste still goes to land fill, i.e. large holes in the ground which are filled with waste products from the home and commercial activities. It is natural that these land-fill sites will leach waste water over the period of their operation which can also act to pollute local rivers and streams. It is natural, therefore, for governments to regulate the use of the land on which we live, work and play to conserve and protect the environment.

Response 1.2

Direct exposure to humans includes inhalation of volatile compounds (e.g. solvents) or dust/soil, absorption through the skin and ingestion (deliberate or unintentional).

Response 1.3

By using the data presented, it is possible to:

(1) Calculate the arithmetic sample mean, $\bar{x}$, as 410.17.

(2) Calculate the (unbiased) sample standard deviation, s, as 93.58.

(3) Obtain the t-value from Table 1.5 for $n = 12$, as $t = 1.796$.

Bioavailability, Bioaccessibility and Mobility of Environmental Contaminants   John R. Dean
© 2007 John Wiley & Sons, Ltd

(4) Calculate the upper 95th percentile bound of the sample as:

$$US_{95} = \bar{x} + \frac{ts}{\sqrt{n}}$$

$$US_{95} = 410.17 + (1.796 \times 93.58)/\sqrt{12}$$

$$US_{95} = 458.69$$

(5) Compare the upper bound value (US_{95}) with the Soil Guideline Value, G.

The upper bound value is lower than the Soil Guideline Value of 500 and therefore it can be concluded that NO action is required in the averaging area based on the mean-value test.

Response 1.4

The objective is to decide whether the maximum value, $x_1 = 520$ ($y_{max} = 2.716$), should be treated as an outlier, or whether it is reasonable to be considered as coming from the same underlying population as the other samples. By using the data presented, it is possible to:

(1) Calculate the arithmetic sample mean of the y-values, as $\bar{y} = 2.600$.

(2) Calculate the (unbiased) standard deviation of the y-values, as $s_y = 0.1206$.

(3) Obtain the outlier test statistic as:

$$T = (y_{max} - \bar{Y})/s_y$$

$$T = (2.716 - 2.600)/0.1206 = 0.96$$

Then, compare this value ($T = 0.96$) with the critical value (T_{crit}) in Table 1.6.

In the context of critical values being used for contaminated soil analysis, the concern is with regard to accepting a value which should be treated as an outlier. As 10% critical values are more stringent, they should be used for health-protection purposes, rather than the 5% values.

Therefore, the maximum value statistic calculated above ($T = 0.96$) is less than the 10% critical value of 2.13. Thus, it is reasonable to conclude that the maximum value is indeed part of the same underlying distribution as the other values.

Chapter 2

Response 2.1

The answer to this question is either very straightforward or not. In one sense, the easiest answer is that you would use the only analytical technique that you have available, whether or not it is 'ideal' for its purpose. However, if you have a range of analytical techniques available, then the decision is more complex and requires further consideration. Issues to consider then include the following:

- What is the metal or POP to be analysed?

- Which technique is the most appropriate for that metal/POP?

- What are the limits of detection that are required for this purpose?

- What precision will be acceptable?

- How will I assess accuracy?

- How fast is the 'turn-around' time for this type of analysis?

- Are analyses needed 'on site'?

- How many samples will there be?

- Is the analysis capable of being automated?

- How much will it all cost? Capital cost of analytical technique? Operating costs?

- Level of skill required by operator? Training of operator?

As I have tried to indicate, the answer to the original question does require some thought. Can you think of any more specific questions?

Response 2.2[†]

Microwaves are high-frequency electromagnetic radiation with a typical wavelength of 1 mm to 1 m. Many microwaves, both industrial and domestic, operate at a wavelength of around 12.2 cm (or a frequency of 2.45 GHz) to prevent interference with radio transmissions. Microwaves are split into two parts, i.e. the electric-field component and the magnetic-field component. These are perpendicular to each other and the direction of propagation (travel) and vary sinusoidally. Microwaves, are comparable to light in their characteristics. They are said to have both particulate character as well as acting like waves. The 'particles' of microwave energy are known as *photons*. These photons are absorbed by the molecule in the lower-energy state (E_0) and the energy raises an electron to a higher-energy level (E_1). Since electrons occupy definite energy levels, changes in these levels are discrete and therefore do not occur continuously. The energy is said to be *quantized*. Only charged particles are affected by the electric-field part of the microwave. The Debye equation for the dielectric constant of a material

---

[†] See Dean, J. R., *Methods for Environmental Trace Analysis*, AnTS Series, John Wiley & Sons, Ltd, Chichester, UK, 2003, pp. 56–57.

determines the polarizability of the molecule. If these charged particles or polar molecules are free to move, this causes a current in the material. However, if they are bound strongly within the compound and consequently are not mobile within the material, a different effect occurs. The particles re-orientate themselves so they are in-phase with the electric field. This is known as *dielectric polarization*.

The latter is split into four components, with each being based upon the four different types of charged particles that are found in matter. These are electrons, nuclei, permanent dipoles and charges at interfaces. The total dielectric polarization of a material is the sum of all four components:

$$\alpha_1 = \alpha_e + \alpha_a + \alpha_d + \alpha_i$$

where α_1 is the total dielectric polarization, α_e is the electronic polarization (polarization of electrons round the nuclei), α_a is the atomic polarization (polarization of the nuclei), α_d is the dipolar polarization (polarization of permanent dipoles in the material) and α_i is the interfacial polarization (polarization of charges at the material interfaces).

The electric field of the microwaves is in a state of flux, i.e. it is continually polarizing and depolarizing. These frequent changes in the electric field of the microwaves cause similar changes in the dielectric polarization. Electronic and atomic polarization and depolarization occur more rapidly than the variation in the electric field and have no effect on the heating of the material. Interfacial polarization (also known as the Maxwell–Wagner effect) only has a significant effect on dielectric heating when charged particles are suspended in a non-conducting medium, and are subjected to microwave radiation. The time-period of oscillation of permanent dipoles is similar to that of the electric field of the microwaves. The resulting polarization lags behind the reversal of the electric field and causes heating in the substance. These phenomena are thought to be the main contributors to dielectric heating.

Response 2.3

A metal complex is where the metal is chemically bound, using lone pairs of electrons, to an organic compound such that the metal is encapsulated or 'chelated'. An example is shown in Figure SAQ 2.3 for ammonium pyrrolidine dithiocarbamate (APDC).

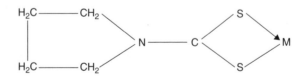

Figure SAQ 2.3 Metal complexation using ammonium pyrrolidine dithiocarbamate (APDC) (note that in this case the ammonium ion has been displaced by a metal (M)).

Response 2.4

Other flames that have been used previously include air–hydrogen (2300 K) and air–propane (2200 K) flames.

Response 2.5

In the Smith–Hieftje background correction approach, a single hollow-cathode lamp (HCL) is used which is operated at high and low (or normal) currents. The operation of an HCL at high-lamp current induces 'self-reversal' in the emission profile of the lamp. It is this feature which is exploited in the Smith–Hieftje background correction. The principle of the method is as follows. The HCL, operating at low current, is able to absorb radiation from both the (desired) atomic and (unwanted) molecular species, as would normally occur. If the HCL is then subjected to a pulse of high current, 'self-reversal' occurs, effectively splitting the profile into two. Incomplete self-reversal in the HCL will lead to incomplete resolving of the atomic line profile from molecular-band interferences. As the atomic species no longer coincide at exactly the same wavelength (as is the case with the HCL at low current), they are not observed, whereas the broad-band absorptions of the molecular species are still observed. Subtraction of the two signals allows a corrected absorption signal to be monitored.

Response 2.6

This refers to three elements and a mineral, namely neodymium (Nd), yttrium (Y), aluminium (Al) and garnet (G).

Response 2.7

These devices are commonly found in digital cameras and camcorders, i.e. those which allow you to visualize objects of a photograph or movie.

Response 2.8

The use of pressure units in science can be quite confusing as often you do not find the SI System (Systeme International d'Unites) being used for historical reasons. Nevertheless, it is very useful to be able to convert between these different units. The SI unit of pressure is the pascal (Pa), which has been defined as $1\,Pa = 1\,kg\,m^{-1}s^{-2} = 1\,N\,m^{-2}$. In addition, other units of pressure that are frequently used include the bar (bar, a cgs unit of pressure), atmosphere (atm), millimetre of mercury (mmHg) and torr (torr):

$$1\,Pa = 9.869\,23 \times 10^{-6}\,atm = 10 \times 10^{-6}\,bar = 7.500\,64$$
$$\times\,10^{-3}\,mmHg = 7.500\,64 \times 10^{-3}\,torr.$$

Therefore, 1 bar = 100 000 Pa. As 2.5 mbar is the same as 2.5×10^{-3} bar, then 2.5×10^{-3} bar = 250 Pa = 2.4673×10^{-3} atm = 1.875 16 mmHg = 1.875 16 torr.

Response 2.9[‡]

The use of collision and reaction cells in ICP–MS allows for the following:

- Neutralization of the most intense chemical ionisation species.

- Interferent or analyte ion mass/charge ratio shifts.

These processes are affected by the use of a range of reaction types, including the following:

- charge exchange

- atom transfer

- adduct formation

- condensation reactions.

General forms of each of these reaction types, with selected examples, are presented in the following (A, analyte; B, reagent; C, interferent):

Charge exchange

$$\text{General form: } C^+ + B \longrightarrow B^+ + C$$

Charge exchange allows the removal of, for example, the argon plasma gas ion interference and the resultant formation of uncharged argon plasma gas which is not then detected.

$$\text{Example: } Ar^+ + NH_3 \longrightarrow NH_3{}^+ + Ar$$

Atom transfer: proton transfer

$$\text{General form: } CH^+ + B \longrightarrow BH^+ + C$$

Proton transfer can remove the interference from, for example, ArH^+. This results in the formation of neutral (uncharged) argon plasma gas which is then not detected.

$$\text{Example: } ArH^+ + H_2 \longrightarrow H_3{}^+ + Ar$$

[‡] See Dean, J. R., *Practical Inductively Coupled Plasma Spectroscopy*, AnTS Series, John Wiley & Sons, Ltd, Chichester, UK, 2005, pp. 110–111.

Atom transfer: hydrogen-atom transfer

$$\text{General form: } C^+ + BH \longrightarrow CH^+ + B$$

Hydrogen-atom transfer has the ability to alleviate an interference by increasing the mass/charge ratio by one.

$$\text{Example: } Ar^+ + H_2 \longrightarrow ArH^+ + H$$

Atom transfer: hydride-atom transfer

$$\text{General form: } A^+ + BH \longrightarrow B^+ + AH$$

Hydride-atom transfer can remove an interference by forming the hydride of an element with no charge.

Adduct formation

$$\text{General form: } A^+ + B \longrightarrow AB^+$$

Adduct formation, with, for example, ammonia, NH_3, allows the mass/charge ratio to increase by 17 amu (atomic weight of $N = 14$ and atomic weight of $H = 1$).

$$\text{Example: } Ni^+ + NH_3 \longrightarrow Ni^{+\cdot}NH_3$$

Condensation reaction

$$\text{General form: } A^+ + BO \longrightarrow AO^+ + B$$

The use of a condensation reaction, in common with adduct formation, has the ability to increase the mass/charge ratio. For example, creation of the oxide of the element will increase the mass/charge ratio by 16 amu (atomic weight of $O = 16$).

$$\text{Example: } Ce^+ + N_2O \longrightarrow CeO^+ + N_2$$

Response 2.10

The identification of element species information or 'speciation' is defined as the process of identifying and quantifying the different defined species, forms or phases present in a material or the description of the amounts and types of these. The reason why speciation is important is that elements can be present in many forms, some of which are toxic. Examples of elements for which toxicity is an issue are arsenic, chromium, mercury and tin.

Arsenic occurs in many chemical forms in the environment, i.e. arsenite (As(III)), arsenate (As(V)), monomethylarsonic acid (MMAA), dimethylarsinic acid (DMAA), arsenobetaine (AsB) and arsenocholine (AsC), with a range of toxicities, e.g. As(III) is the most toxic form of arsenic. In fish tissue, the main form of arsenic is arsenobetaine, which is non-toxic. Trivalent chromium (Cr(III)) is an essential element for man as it is involved in glucose, lipid and protein metabolism. In contrast, hexavalent chromium (Cr(VI)) is a potent carcinogen. It is therefore essential to be able to distinguish between these two different oxidation states of chromium. All forms of mercury are considered to be poisonous. However, it is methylmercury (or as the chloride, CH_3HgCl) which is considered to be the most toxic because of its ability to bioaccumulate in fish. The best example of the toxicity of methylmercury occurred in Minamata in Japan in 1955. It was found that methylmercury-contaminated fish consumed by pregnant women resulted in the 'new-born' children having severe brain damage (*Minamata disease*). Butyl- and phenyltin compounds are known to be toxic in the marine environment. Historically, tributyltin (TBT) has been released into the marine environment from the leaching of TBT-based antifouling paints used on the undersides of boats and ships. Triphenyltin (TPhT) has also been used as an antifouling agent in paint and in herbicide formulations.

Chapter 3

Response 3.1

Extraction of POPs from solid or semi-solid matrices can be divided into several categories based on the method of extraction, mode of heating and presence (or not) of some type of agitation (Figure SAQ 3.1).

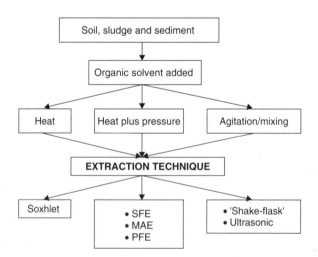

Figure SAQ 3.1 Procedure adopted in the extraction of persistent organic pollutants from (semi)-solid matrices.

Response 3.2[†]

Microwaves are high-frequency electromagnetic radiation with a typical wavelength of 1 mm to 1 m. Many microwaves, both industrial and domestic, operate at a wavelength of around 12.2 cm (or a frequency of 2.45 GHz) to prevent interference with radio transmissions. Microwaves are split into two parts, i.e. the electric-field component and the magnetic-field component. These are perpendicular to each other and the direction of propagation (travel) and vary sinusoidally. Microwaves, are comparable to light in their characteristics. They are said to have both particulate character as well as acting like waves. The 'particles' of microwave energy are known as *photons*. These photons are absorbed by the molecule in the lower-energy state (E_0) and the energy raises an electron to a higher-energy level (E_1). Since electrons occupy definite energy levels, changes in these levels are discrete and therefore do not occur continuously. The energy is said to be *quantized*. Only charged particles are affected by the electric-field part of the microwave. The Debye equation for the dielectric constant of a material determines the polarizability of the molecule. If these charged particles or polar molecules are free to move, this causes a current in the material. However, if they are bound strongly within the compound and consequently are not mobile within the material, a different effect occurs. The particles re-orientate themselves so they are in-phase with the electric field. This is known as *dielectric polarization*.

Response 3.3[‡]

Liquid solvents at elevated temperatures and pressures should provide enhanced extraction capabilities compared to their use at or near room temperature and atmospheric pressure for two main reasons: (i) solubility and mass transfer effects and (ii) disruption of surface equilibria.

Solubility and Mass-Transfer Effects
The following three factors are considered important:

- Higher temperatures increase the capacity of solvents to solubilize analytes.

- Faster diffusion rates occur as a result of increased temperatures.

- Improved mass transfer, and hence increased extraction rates, occur when fresh solvent is introduced, i.e. the concentration gradient is greater between the fresh solvent and the surface of the sample matrix.

Disruption of Surface Equilibria
As both temperature and pressure are important, both are discussed separately.

[†] See Dean, J. R., *Methods for Environmental Trace Analysis*, AnTS Series, John Wiley & Sons, Ltd, Chichester, UK, 2003, pp. 56–57.
[‡] See Dean, J. R., *Methods for Environmental Trace Analysis*, AnTS Series, John Wiley & Sons, Ltd, Chichester, UK, 2003, pp. 129–130.

Temperature Effects

- Increased temperatures can disrupt the strong solute–matrix interactions caused by van der Waals forces, hydrogen bonding and dipole attractions of the solute molecules and active sites on the matrix.

- Decreases in the viscosities and surface tensions of solvents occur at higher temperatures, thus allowing improved penetration of the matrix, and hence improved extraction.

Pressure Effects

- The utilization of elevated pressures allows solvents to remain liquified above their boiling points.

- Extraction from within the matrix is possible, as the pressure allows the solvent to penetrate the sample matrix.

Response 3.4

Important points to consider when comparing different extraction techniques for the recovery of POPs from solid matrices include the following:

- How much sample is required?
- How long will the extraction take?
- Are there any limitations on the choice of solvents?
- How much solvent is required?
- Can the extraction be done sequentially or simultaneously?
- Is it possible to automate the process?
- Will it require low/medium/high operator skills to perform the extractions?
- What is the capital cost of the apparatus/equipment?
- What are the consumable costs?
- Does the extraction process have any approved methods?
- Is the extraction system commercially available?

Response 3.5[§]

Two terms are used to describe the distribution of a POP between two immiscible solvents, i.e. the *distribution coefficient* and the *distribution ratio*. The distribution

---

[§] See Dean, J. R., *Methods for Environmental Trace Analysis*, AnTS Series, John Wiley & Sons, Ltd, Chichester, UK, 2003, pp. 100–101.

coefficient is an equilibrium constant which describes the distribution of a POP, between two immiscible solvents, e.g. an aqueous and an organic phase. For example, an equilibrium can be obtained by shaking the aqueous phase containing the POP with an organic phase, such as hexane. This process can be written as an equation, as follows:

$$POP(aq) \rightleftharpoons POP(org) \qquad \text{(SAQ 1)}$$

where (aq) and (org) represent the aqueous and organic phases, respectively. The ratio of the activities of the POP in the two solvents is constant and can be represented by the following:

$$K_d = [POP]_{org}/[POP]_{aq} \qquad \text{(SAQ 2)}$$

where K_d is the distribution coefficient. While the numerical value of K_d provides a useful constant value, at a particular temperature, the activity coefficients are neither known or easily measured. A more useful expression is the fraction of POP extracted, E, often expressed as a percentage, as follows:

$$E = C_o V_o/(C_o V_o + C_{aq} V_{aq}) \qquad \text{(SAQ 3)}$$

or:

$$E = K_d V/(1 + K_d V) \qquad \text{(SAQ 4)}$$

where C_o and C_{aq} are the concentrations of the POP in the organic and aqueous phases, respectively, V_o and V_{aq} are the volumes of the organic and aqueous phases, respectively, and V is the phase ratio, i.e. V_o/V_{aq}.

For one-step liquid–liquid extractions, K_d must be large, i.e. > 10, for quantitative recovery (> 99%) of the POP in one of the phases, e.g. the organic solvent. This is a consequence of the phase ratio, V, which must be maintained within a practical range of values: $0.1 < V < 10$ (equation (SAQ 4)). Typically, two or three repeat extractions are required with fresh organic solvent to achieve quantitative recoveries. Equation (SAQ 5) is used to determine the amount of POP extracted after successive multiple extractions:

$$E = 1 - [1/(1 + K_d V)]^n \qquad \text{(SAQ 5)}$$

where n is the number of extractions. For example, if the volume of the two phases are equal ($V = 1$) and $K_d = 3$ for an analyte, then four extractions ($n = 4$) would be required to achieve > 99% recovery.

Response 3.6

The purpose of SPE is to recover from aqueous solution, using a solid support, any POPs that might be present. Then, suitable washing of the solid support by a small volume of organic solvent allows easy release of the retained POPs. In this manner, the POPs present in a large volume of aqueous solution are pre-concentrated prior to analysis. An alternative approach for SPE is to use the technique to 'clean-up' a sample, i.e. to remove POPs from any extraneous material.

Response 3.7

Important aspects to consider when comparing different extraction techniques for the recovery of POPs from liquid samples include the following:

- How much sample is required?

- How long will the extraction take?

- Are there any limitations on the choice of solvents?

- How much solvent is required?

- Can the extraction be done sequentially or simultaneously?

- Is it possible to automate the process?

- Will it require low/medium/high operator skills to perform the extractions?

- What is the capital cost of the apparatus/equipment?

- What are the consumable costs?

- Does the extraction process have any approved methods?

- Is the extraction system commercially available?

Chapter 4

Response 4.1

If the nebulizer does become blocked, the following cleaning procedures can be used:

- Remove the nebulizer from its mounting and visually examine under 20× or 30× magnification.

- Particle wedged inside nebulizer.

  - Carefully tap the liquid input of the nebulizer gently. If the particle becomes loose, repeat the tapping to allow the particle to progress to the exit orifice.

  - Apply compressed gas (up to 30 psi) to the nozzle.

  - 'Back-flush' the nozzle using isopropyl alcohol.

• Solid material present but a flow through the nebulizer is still possible.

  - Inject appropriate solvent into the nozzle to dissolve solid deposit.

  - Remove the solvent with compressed gas.

  - Repeat above, but gently heat the nebulizer.

• Solid material present and nebulizer is blocked.

  - Gently heat the nebulizer at the point of the blockage. Then carefully apply gas pressure at the sample input tube.

• Nozzle is encrusted with crystalline deposits.

  - Immerse the nozzle in an appropriate rinse solution.

  - Apply compressed gas (up to 30 psi) to the nozzle.

• No foreign matter is visible.

  - Immerse the nozzle in hot, concentrated nitric acid. Repeat as necessary.

Response 4.2
Aqua regia is prepared from concentrated hydrochloric acid (HCl) and nitric acid (HNO_3) in the following ratio, 3:1 (vol/vol). This means, for example, placing 75 ml of hydrochloric acid and 25 ml of nitric acid into a 100 ml measuring cylinder.

Response 4.3
Earthworms travel underground by means of a series of waves of muscular contractions which alternately shorten and lengthen the body [11]. In the shortened state, the earthworm is anchored to the surrounding soil by tiny claw-like bristles, often called setae, which are distributed along its segmented length. Normally, earthworms have four pairs of setae per segment but some genera are perichaetin, i.e. they have a larger number of setae on each segment. The entire movement process is assisted by the earthworm secreting a lubricating mucus.

Response 4.4
Diethylenetriamine pentaacetic acid.

Response 4.5
(a) It is noted that the numerical values are often grouped so that the numbers are within 1000 of each other with the appropriate choice of units, e.g.

$0.050\,\mu g\,g^{-1}$ selenium is within a factor of 1000 for rubidium ($10.2\,\mu g\,g^{-1}$) but not within a factor of 1000 for magnesium (0.271 wt% or $2710\,\mu g\,g^{-1}$). In addition, for the major constituents, two or three significant figures are quoted for values in the wt% range. This allows large concentrations to be quoted without the necessity to report excessive significant figures, e.g. instead of quoting $16\,100\,\mu g\,g^{-1}$ for potassium it is more appropriate to quote 1.61 wt%. All values are quoted with a variation ($\pm$) of one standard deviation (SD) of the mean value.

(b) In order to allow a direct comparison from one laboratory to another, some sort of normalization is required. In this case, it is 'dry weight'. As indicated in the footnote, a sub-sample of the SRM needs to be dried under specific conditions and for a specific time. Following this exact procedure allows any user of the SRM the opportunity to obtain comparable data. Users of the SRM then prepare and analyse the material using their own procedures and compare their data directly with those on the certificate. If data are obtained within the mean $\pm$ one standard deviation (SD), then the 'in-house' procedure is appropriate. If, however, the data on the certificate cannot be replicated, this is perhaps indicative of poor 'in-house' procedures and thus warrants further investigation.

Chapter 5

Response 5.1

If you have not already done so, you should read Chapter 3 first which outlines the different extraction techniques. In summary, however, the techniques used include the use of pressurized fluid extraction (PFE), microwave-assisted extraction (MAE) and Soxhlet extraction[†] for the recovery of POPs from environmental matrices employing polar organic solvents (dichloromethane, methanol, etc.), coupled with either pressure and/or temperature to effectively release the analytes from (semi)-solid matrices (soil, sludge, etc.).

Response 5.2

Organic solvents can be classified in terms of their polarity by using the dielectric constants[‡]. Table 5.1 indicates the different polarities of a range of common organic solvents used for extracting POPs from solid matrices. For completeness, a range of other solvent properties is included.

[†] See Dean, J. R., *Methods for Environmental Trace Analysis*, AnTS Series, John Wiley & Sons, Ltd, Chichester, UK, 2003.
[‡] See Lide, D. R. (Ed.), *CRC Handbook of Chemistry and Physics*, 73rd Edition, CRC Press, Boca Raton, FL, USA, 1992–1993.

Response 5.3

The total Hildebrand solubility parameter, δ_t, for acetonitrile is 25.0, while the δ_t for dichloromethane is 20.5.

Response 5.4

The total Hildebrand solubility parameters, δ_t, for lindane and hexachlorobenzene are as follows: δ_t for lindane is 27.1, while δ_t for hexachlorobenzene is 23.6.

Response 5.5

For a 1:4 (vol/vol) ratio of acetonitrile and dichloromethane, the dispersion coefficient is as follows: $\delta_d = (0.2\delta_{d(acetonitrile)} + 0.4\delta_{d(dichloromethane)})$.

Response 5.6

Using the Hildebrand solubility parameter approach, toluene would be predicted to be a weak solvent for the recovery of lindane, endosulfan I, p, p'-DDE, endrin, and p, p'-DDD, while acetone would be predicted to be a strong solvent.

Response 5.7

In the case of lindane (Figure 5.11(a)), the total recoveries by both solvents $(8.8\,mg\,kg^{-1})$ were in agreement with both the certificate value and the recovery using a single solvent (acetone) only (Table 5.7). In the case of endosulfan I (Figure 5.11(b)), the total recoveries by both solvents $(4.5\,mg\,kg^{-1})$ were in agreement with the certificate value but lower than that obtained by a single solvent (acetone) only (Table 5.7). In the case of p, p'-DDE (Figure 5.11(c)), the total recoveries by both solvents $(11.8\,mg\,kg^{-1})$ were in agreement with both the certificate value and the recovery using a single solvent (acetone) only (Table 5.7). In the case of endrin (Figure 5.11(d)), the total recoveries by both solvents $(10.7\,mg\,kg^{-1})$ were in agreement with the certificate value but lower than that obtained by a single solvent (acetone) only (Table 5.7). Finally, In the case of p, p'-DDD (Figure 5.11(e)), the total recoveries by both solvents $(19.2\,mg\,kg^{-1})$ were in agreement with both the certificate value and the recovery using a single solvent (acetone) only (Table 5.7).

Response 5.8

Linear regression analysis uses an equation of the following form:

$$Y = mx + c$$

where Y is the signal obtained, m is the 'slope', x is the concentration (of the variable) and c is the intercept of the line on the x-axis.

Chapter 6

Response 6.1

The unintentional consumption of soil may arise, for example, from unwashed foodstuffs, including vegetables or poor personal hygiene, e.g. small children putting their fingers in their mouths after playing in the garden.

Response 6.2

The seven specialist functions of the gastrointestinal tract are as follows:

- *Ingestion* – the intake of food into the digestive system via the mouth.

- *Mastication* – the pulverization of foodstuffs by the action of chewing in the presence of saliva.

- *Deglutition* – the action of swallowing of the food, allowing it to pass from the mouth to the stomach.

- *Digestion* – the mechanical and chemical breakdown of foodstuffs in the stomach.

- *Absorption* – the transfer of foodstuffs through the mucous membrane of the intestines into the blood stream.

- *Peristalsis* – the rhythmic, wavelike contractions that move food through the digestive tract.

- *Defecation* – the discharge of undigestible waste products from the body.

Response 6.3

The peristaltic actions of the oesophagus can best be mimicked in the laboratory by shaking or agitation of the sample. This can be achieved by using, for example, an 'end-over-end' shaker.

Glossary of Terms

This section contains a glossary of terms, all of which are used in the text. It is not intended to be exhaustive, but to explain briefly those terms which often cause difficulties or may be confusing to the inexperienced reader.

Accelerated solvent extraction (ASE) Method of extracting analytes from matrices using a solvent at elevated pressure and temperature (*see also* Pressurized fluid extraction).

Accuracy A quantity referring to the difference between the mean of a set of results or an individual result and the value which is accepted as the true or correct value for the quantity measured.

Acid digestion Use of acid (and often heat) to destroy the organic matrix of a sample to liberate the metal content.

Aliquot A known amount of a homogenous material assumed to be taken with negligible sampling error.

Analyte The component of a sample which is ultimately determined directly or indirectly.

Calibration The set of operations which establish, under specified conditions, the relationship between values indicated by a measuring instrument or measuring system and the corresponding known values of the measurand.

Calibration curve Graphical representation of the measuring signal as a function of the quantity of analyte.

Cation An ion having a positive charge. Atoms of metals, in solution, become cations.

Bioavailability, Bioaccessibility and Mobility of Environmental Contaminants   John R. Dean
© 2007 John Wiley & Sons, Ltd

Certified Reference Material (CRM) A reference material, accompanied by a certificate, one or more of whose property values are certified by a procedure which establishes its traceability to an accurate realization of the unit in which the property values are expressed, and for which each certified value is accompanied by an uncertainty at a stated level of confidence.

Complexing agent The chemical species (an ion or a compound) which will bond to a metal ion using lone pairs of electrons.

Confidence interval Range of values which contains the true value at a given level of probability. The level of probability is called the *confidence level*.

Confidence limit The extreme values or end values in a confidence interval.

Contamination In trace analysis, this is the unintentional introduction of analyte(s) or other species which are not present in the original sample and which may cause an error in the determination. It can occur at any stage in the analysis. Quality assurance procedures, such as analyses of blanks or of reference materials, are used to check for contamination problems.

Control of Substances Hazardous to Health (COSHH) Regulations which impose specific legal requirements for risk assessment wherever hazardous chemicals or biological agents are used.

Co-precipitation The inclusion of otherwise soluble ions during the precipitation of lower-solubility species.

Dissolved material Refers to material which will pass through a 0.45 μm membrane filter assembly prior to sample acidification.

Dry ashing Use of heat to destroy the organic matrix of a sample to liberate the metal content.

Extraction The removal of a soluble material from a solid mixture by means of a solvent, or the removal of one or more components from a liquid mixture by use of a solvent with which the liquid is immiscible or nearly so.

Heterogeneity The degree to which a property or a constituent is randomly distributed throughout a quantity of material. The degree of heterogeneity is the determining factor of sampling error.

Homogeneity The degree to which a property or a constituent is uniformly distributed throughout a quantity of material. A material may be homogenous with respect to one analyte but heterogeneous with respect to another.

Humic acid Naturally occurring high-molecular-mass organic compounds which are acid-soluble but are precipitated by base.

Interferent Any component of the sample affecting the final measurement.

Limit of detection The detection limit of an individual analytical procedure is the lowest amount of an analyte in a sample which can be detected but not necessarily quantified as an exact value. The limit of detection expressed as the concentration, c_L, or the quantity, q_L, is derived from the smallest measure, x_L, which can be detected with reasonable certainty for a given procedure. The value of x_L is given by the following equation:

$$x_L = x_{bl} + k s_{bl}$$

where x_{bl} is the mean of the blank measures, s_{bl} is the standard deviation of the blank measures and k is a numerical factor chosen according to the confidence level required. For many purposes the limit of detection is taken to be $3 s_{bl}$ or $3 \times$ 'the signal-to-noise ratio', assuming a 'zero-blank.'

Limit of quantitation For an individual analytical procedure, this is the lowest amount of an analyte in a sample which can be quantitatively determined with suitable uncertainty. It may also be referred to as the *limit of determination*. The limit of quantitation can be taken as $10 \times$ 'the signal-to-noise ratio', assuming a 'zero blank'.

Linearity Defines the ability of the method to obtain test results proportional to the concentration of analyte.

Liquid–liquid extraction A method of extracting a desired component from a liquid mixture by bringing the solution into contact with a second liquid, the solvent, in which the component is also soluble, and which is immiscible with the first liquid, or nearly so.

Matrix The carrier of the test component (analyte): all the constituents of the material except the analyte, or the material with as low a concentration of the analyte as it is possible to obtain.

Method The overall, systematic procedure required to undertaken an analysis. It includes all stages of the analysis – not just the (instrumental) end determination.

Microwave-assisted extraction (MAE) Method of extracting analytes from matrices using solvent at elevated temperature (and pressure) based on microwave radiation. Can be carried out in open or sealed vessels.

Microwave digestion Method of digesting an organic matrix to liberate metal content using acid at elevated temperature (and pressure) based on microwave radiation. Can be carried out in open or sealed vessels.

Outlier An observation in a set of data which appears to be inconsistent with the remainder of that set.

Pesticide Any substance or mixture of substances intended for preventing, destroying, repelling or mitigating any pest. The latter can be insects, mice and other animals, unwanted plants (weeds), fungi or micro-organisms, such as bacteria and viruses. Although often misunderstood to refer only to *insecticides*, the term 'pesticide' also applies to herbicides, fungicides and various other substances used to control pests.

Polycyclic aromatic hydrocarbons (PAHs) These are a large group of organic compounds, comprising two or more aromatic rings, which are widely distributed in the environment.

Precision The closeness of agreement between independent test results obtained under stipulated conditions.

Pressurized fluid extraction (PFE) Method of extracting analytes from matrices using solvent at elevated pressures and temperatures (*see also* Accelerated solvent extraction).

Qualitative analysis Chemical analysis designed to identify the components of a substance or mixture.

Quantitative analysis Chemical analysis which is normally taken to mean the numerical measurement of one or more analytes to the required level of confidence.

Reagent A test substance that is added to a system in order to bring about a reaction or to see whether a reaction occurs (e.g. an analytical reagent).

Reagent blank A solution obtained by carrying out all of the steps of the analytical procedure in the absence of a sample.

Recovery The fraction of the total quantity of a substance recoverable following a chemical procedure.

Reference material A substance or material, one or more of whose property values are sufficiently homogeneous and well-established to be used for the calibration of an apparatus, the assessment of a measurement method or for assigning values to materials.

Repeatability Precision under repeatability conditions, i.e. conditions where independent test results are obtained with the same method on identical test items in the same laboratory, by the same operator, using the same equipment within short intervals of time.

Reproducibility Precision under reproducibility conditions, i.e. conditions where test results are obtained with the same method on identical test items in different laboratories, with different operators, using different equipment.

Robustness A measure of the capacity of an analytical procedure to remain unaffected by small, but deliberate variations in method parameters, and which provides an indication of its reliability during normal usage. Sometimes referred to as *ruggedness*.

Rotary evaporation Removal of solvents by distillation under vacuum.

Sample A portion of material selected from a larger quantity of material. The term needs to be qualified, e.g. representative sample, sub-sample, etc.

Selectivity (in analysis) (i) *Qualitative* – the extent to which other substances interfere with the determination of a substance according to a given procedure. (ii) *Quantitative* – a term used in conjunction with another substantive (e.g. constant, coefficient, index, factor, number, etc.) for the quantitative characterization of interferences.

Sensitivity The change in the response of a measuring instrument divided by the corresponding change in stimulus.

Signal-to-noise ratio A measure of the relative influence of noise on a control signal. Usually taken as the magnitude of the signal divided by the standard deviation of the background signal.

'Shake-flask' extraction Method of extracting analytes from matrices using agitation or shaking in the presence of a solvent.

Solid-phase extraction (SPE) A sample preparation technique which uses a solid-phase packing contained in a small plastic cartridge. The solid stationary phases are the same as HPLC packings; however, the principle is different from HPLC. The process, as most often practiced, requires four steps: conditioning the sorbent, adding the sample, washing away the impurities and eluting the sample in as small a volume as possible with a strong solvent.

Solid-phase microextraction (SPME) A sample preparation technique which uses a fused silica fibre coated with a polymeric phase to sample either an aqueous solution or the headspace above a sample. Analytes are absorbed by the polymer coating and the SPME fiber is directly transferred to a GC injector or special HPLC injector for desorption and analysis.

Solvent extraction The removal of a soluble material from a solid mixture by means of a solvent, or the removal of one or more components from a liquid mixture by use of a solvent with which the liquid is immiscible, or nearly so.

Soxhlet extraction Equipment for the continuous extraction of a solid by a solvent. The material to be extracted is placed in a porous cellulose thimble and continually condensing solvent allowed to percolate through it and return to the boiling vessel, either continuously or intermittently.

Speciation The process of identifying and quantifying the different defined species, forms or phases present in a material or the description of the amounts and types of these species, forms or phases present.

Standard (general) An entity established by consensus and approved by a recognized body. It may refer to a material or solution (e.g. an organic compound of known purity or an aqueous solution of a metal of agreed concentration) or a document (e.g. a methodology for an analysis or a quality system). The relevant terms are as follows:

Analytical standard (*also known as* **Standard solution**) A solution or matrix containing the analyte which will be used to check the performance of the method/instrument.

Calibration standard The solution or matrix containing the analyte (measurand) at a known value with which to establish a corresponding response from the method/instrument.

External standard A measurand, usually identical with the analyte, which is analysed separately from the sample.

Internal standard A measurand, similar to but not identical with the analyte, which is combined with the sample.

Standard method A procedure for carrying out a chemical analysis which has been documented and approved by a recognized body.

Standard addition The addition of a known amount of analyte to a sample in order to determine the relative response of the detector to an analyte within the sample matrix. The relative response is then used to assess the sample analyte concentration.

Stock solution This is generally a standard or reagent solution of known accepted stability, which has been prepared in relatively large amounts, of which portions are used as required. Frequently, such portions are used following further dilution.

Sub-sample This may be: (i) a portion of the sample obtained by selection or division, (ii) an individual unit of the lot taken as part of the sample or (iii) the final unit of multi-stage sampling.

Supercritical fluid extraction (SFE) A method of extracting analytes from matrices using a supercritical fluid at elevated pressure and temperature. The term *supercritical fluid* is used to describe any substance above its critical temperature and critical pressure.

True value A value consistent with the definition of a given particular quantity.

Ultrasonic extraction A method of extracting analytes from matrices with solvent, using either an ultrasonic bath or probe.

SI Units and Physical Constants

SI Units

The SI system of units is generally used throughout this book. It should be noted, however, that according to present practice, there are some exceptions to this, for example, wavenumber (cm^{-1}) and ionization energy (eV).

Base SI units and physical quantities

Quantity	Symbol	SI Unit	Symbol
length	l	metre	m
mass	m	kilogram	kg
time	t	second	s
electric current	I	ampere	A
thermodynamic temperature	T	kelvin	K
amount of substance	n	mole	mol
luminous intensity	I_v	candela	cd

Prefixes used for SI units

Factor	Prefix	Symbol
10^{21}	zetta	Z
10^{18}	exa	E
10^{15}	peta	P
10^{12}	tera	T

(continued overleaf)

Bioavailability, Bioaccessibility and Mobility of Environmental Contaminants   John R. Dean
© 2007 John Wiley & Sons, Ltd

Prefixes used for SI units (*continued*)

Factor	Prefix	Symbol
10^9	giga	G
10^6	mega	M
10^3	kilo	k
10^2	hecto	h
10	deca	da
10^{-1}	deci	d
10^{-2}	centi	c
10^{-3}	milli	m
10^{-6}	micro	μ
10^{-9}	nano	n
10^{-12}	pico	p
10^{-15}	femto	f
10^{-18}	atto	a
10^{-21}	zepto	z

Derived SI units with special names and symbols

Physical quantity	SI unit		Expression in terms of base or derived SI units
	Name	Symbol	
frequency	hertz	Hz	$1\,Hz = 1\,s^{-1}$
force	newton	N	$1\,N = 1\,kg\,m\,s^{-2}$
pressure; stress	pascal	Pa	$1\,Pa = 1\,Nm^{-2}$
energy; work; quantity of heat	joule	J	$1\,J = 1\,Nm$
power	watt	W	$1\,W = 1\,J\,s^{-1}$
electric charge; quantity of electricity	coulomb	C	$1\,C = 1\,A\,s$
electric potential; potential difference; electromotive force; tension	volt	V	$1\,V = 1\,J\,C^{-1}$
electric capacitance	farad	F	$1\,F = 1\,C\,V^{-1}$
electric resistance	ohm	Ω	$1\,\Omega = 1\,V\,A^{-1}$
electric conductance	siemens	S	$1\,S = 1\,\Omega^{-1}$
magnetic flux; flux of magnetic induction	Weber	Wb	$1\,Wb = 1\,V\,s$
magnetic flux density;	tesla	T	$1\,T = 1\,Wb\,m^{-2}$
magnetic induction inductance	henry	H	$1\,H = 1\,Wb\,A^{-1}$

Derived SI units with special names and symbols (*continued*)

Physical quantity	SI unit		Expression in terms of base or derived SI units
	Name	Symbol	
Celsius temperature	degree Celsius	°C	$1\,°C = 1\ K$
luminous flux	lumen	lm	$1\ lm = 1\ cd\ sr$
illuminance	lux	lx	$1\ lx = 1\ lm\ m^{-2}$
activity (of a radionuclide)	becquerel	Bq	$1\ Bq = 1\ s^{-1}$
absorbed dose; specific energy	gray	Gy	$1\ Gy = 1\ J\ kg^{-1}$
dose equivalent	sievert	Sv	$1\ Sv = 1\ J\ kg^{-1}$
plane angle	radian	rad	1^a
solid angle	steradian	sr	1^a

[a] rad and sr may be included or omitted in expressions for the derived units.

Physical Constants

Recommended values of selected physical constants[a]

Constant	Symbol	Value
acceleration of free fall (acceleration due to gravity)	g_n	$9.806\,65\ ms^{-2}$ [b]
atomic mass constant (unified atomic mass unit)	m_u	$1.660\,540\,2(10) \times 10^{-27}\ kg$
Avogadro constant	L, N_A	$6.022\,136\,7(36) \times 10^{23}\ mol^{-1}$
Boltzmann constant	k_B	$1.380\,658(12) \times 10^{-23}\ J\ K^{-1}$
electron specific charge (charge-to-mass ratio)	$-e/m_e$	$-1.758\,819 \times 10^{11}\ Ckg^{-1}$
electron charge (elementary charge)	e	$1.602\,177\,33(49) \times 10^{-19}\ C$
Faraday constant	F	$9.648\,530\,9(29) \times 10^4\ C\ mol^{-1}$
ice-point temperature	T_{ice}	$273.15\ K$ [b]
molar gas constant	R	$8.314\,510(70)\ JK^{-1}\ mol^{-1}$
molar volume of ideal gas (at 273.15 K and 101 325 Pa)	V_m	$22.414\,10(19) \times 10^{-3}\ m^3\ mol^{-1}$
Planck constant	h	$6.626\,075\,5(40) \times 10^{-34}\ J\ s$
standard atmosphere	atm	$101\,325\ Pa$ [b]
speed of light in vacuum	c	$2.997\,924\,58 \times 10^8\ ms^{-1}$ [b]

[a] Data are presented in their full precision, although often no more than the first four or five significant digits are used; figures in parentheses represent the standard deviation uncertainty in the least significant digits.
[b] Exactly defined values.

The Periodic Table

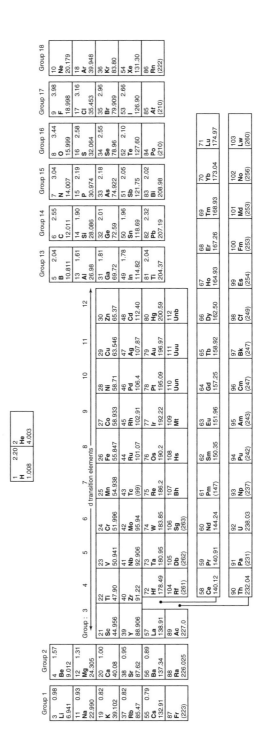

Index